Residenti

Streets

THIRD EDITION

Walter M. Kulash
Principal Author

ULI–the Urban Land Institute
National Association of Home Builders
American Society of Civil Engineers
Institute of Transportation Engineers

ULI–the Urban Land Institute

ULI–the Urban Land Institute is a nonprofit education and research institute that is supported by its members. Its mission is to provide responsible leadership in the use of land in order to enhance the total environment.

ULI sponsors education programs and forums to encourage an open international exchange of ideas and sharing of experiences; initiates research that anticipates emerging land use trends and issues and proposes creative solutions based on that research; provides advisory services; and publishes a wide variety of materials to disseminate information on land use and development. Established in 1936, the Institute today has more than 16,000 members and associates from more than 50 countries representing the entire spectrum of the land use and development disciplines.

Richard M. Rosan
President

ASCE

Founded in 1852, ASCE has a membership of more than 123,000 civil engineers worldwide, and is America's oldest national engineering society. Its mission is to advance professional knowledge and improve the practice of civil engineering.

ITE

The Institute of Transportation Engineers (ITE), an international individual-member educational and scientific association, is one of the largest and fastest-growing multimodal professional transportation organizations in the world. ITE members are traffic engineers, transportation planners and other professionals who are responsible for meeting society's needs for safe and efficient surface transportation through planning, designing, implementing, operating, and maintaining surface transportation systems worldwide.

NAHB

The national Association of Home Builders (NAHB) is a Washington, D.C.–based trade association representing more than 203,000 member firms and professionals involved in homebuilding, remodeling, multifamily construction, property management, subcontracting, design, housing finance, building product manufacturing, and other aspects of residential and light commercial construction. Known as "the voice of the housing industry," NAHB is affiliated with more than 800 state and local homebuilders associations around the country. NAHB's builder members will construct about 80 percent of the more than 1.5 million new housing units projected for 2001.

ULI Project Staff

Rachelle L. Levitt
Senior Vice President, Policy and Practice
Publisher

Gayle Berens
Vice President, Real Estate Development Practice

Robert T. Dunphy
Senior Resident Fellow, Transportation
Project Director

Leslie Holst
Research Associate

Nancy H. Stewart
Director, Book Program

Libby Howland
Manuscript Editor

Meg Batdorff
Graphic Designer
Book/Cover Design, Layout

Diann Stanley-Austin
Director, Publishing Operations

Recommended bibliographic listing:
National Association of Home Builders, American Society of Civil Engineers, Institute of Transportation Engineers, and Urban Land Institute. *Residential Streets,* Third Edition. Washington, D.C.: ULI–the Urban Land Institute, 2001.

ULI Catalog Number: R34
International Standard Book Number: 0-87420-879-3
Library of Congress Control Number: 2001090213

1025 Thomas Jefferson Street, N.W.
Suite 500 West
Washington, D.C. 20007-5201

Second Printing, 2002.

Cover photograph: Vermillion, Huntersville, North Carolina; ©Duany/Plater-Zyberk; Photographer: Thomas E. Low

Acknowledgments

This edition of *Residential Streets* was developed by the National Association of Home Builders, the American Society of Civil Engineers, the Institute of Transportation Engineers, and ULI–the Urban Land Institute. The partners thank the following individuals for their overall guidance and contributions:

William R. Eager
TDA Inc.
Seattle, Washington

Robert E. Engstrom
Robert Engstrom Companies
Minneapolis, Minnesota

David R. Jensen
David Jensen Associates, Inc.
Denver, Colorado

Anne Vernez Moudon
University of Washington
Seattle, Washington

Joe W. Ruffer, PE
Mobile County Department of
Public Works
Mobile, Alabama

Representatives of the partners who directed this work were:

NAHB
John McHenry, AICP, Land Use Planner
Robert K. McNamara, AICP, Senior Policy Planner

ASCE
Amar Chaker, Senior Manager, Technical Activities

ITE
Thomas W. Brahms, Executive Director

ULI
Robert T. Dunphy, Senior Resident Fellow, Transportation

Walter M. Kulash, PE, principal of Glatting Jackson Kercher Anglin Lopez Rinehart Inc. under contract to the partnership, revised the previous edition and prepared additional material. The firm also prepared a number of new graphics for this edition. David Jensen provided information on trends in street planning and community design for the Foreword, and members of his firm helped in providing photographs and plans. TDA Inc. and the Mobile County Department of Public Works, especially James Foster, also provided additional information and photos. Members of the Local Roads and Streets Committee of ASCE's Highway Division provided information and design considerations.

Thanks go to ULI staff Gayle Berens, Bob Dunphy, and Leslie Holst for developing the manuscript and artwork. Thanks also go to the ULI production staff, in particular to Nancy Stewart, Libby Howland, Meg Batdorff, and Diann Stanley-Austin.

Contents

Acknowledgments iii

Foreword ix

CHAPTER 1: Introduction 1

History 1

Early Standard Setting 3

More Recent Efforts 4

Philosophy of Residential Street Design 6

Public and Private Streets 8

Applications 8

CHAPTER 2: Design Considerations 9

Functional Classification System for Streets 9

Types of Traffic Flow 13

Emergency Vehicle Access 14

Traffic Volumes and Residential Dwelling Units 15

Neighborhood Access to Adjacent Street System 17

Street Layout and Platting 18

Principles of Street Layout 20

Number of Lanes 21

Pavement Widths 22

Right-of-Way Widths 25

On-Street Parking 26

Off-Street Parking 27

Alleys 28

Design Speed 29

Gradients 30

Vertical and Horizontal Alignments 31

Dead-End Turnarounds 32

Shared Driveways 36

Streetscape 37

Pedestrian and Bicycle Access 40

Curb Cut Ramps 44

Curbs 46

Traffic Calming 49

CHAPTER 3: Intersections 51
Geometry of Intersections 52
Intersection Angle 53
Traffic Circles and Roundabouts 53
Intersection Spacing 55
Curb Radius 56
Vertical Alignment on Intersection Approaches 57
Corner Sight Distance 58

CHAPTER 4: Streets as Drainage System 59
Closed and Open Systems 59
Street and Curb Cross-Sections 62
Runoff Amounts 63
Criteria for the Spread of Water across Streets 63
Flow across Intersections 64
Ponding at Low Points in Grade 64
Maximum Velocity in Gutters 64
Block and Lot Grading 65
Cut-and-Fill Embankment Slopes 66

CHAPTER 5: Pavement 67
Subgrade 69
Base 69
Wearing Surface 70
Life-Cycle Cost Analysis 72

References 73

Index 75

Figures

Chapter 1: Introduction

1-1 Early postwar streetscapes . . . 2
1-2 Street design as art . . . 4
1-3 Italian street scene . . . 5
1-4 Bike path . . . 6

Chapter 2: Design Considerations

2-1 Hierarchy of streets . . . 10
2-2 Freeway . . . 10
2-3 Arterial street . . . 11
2-4 Collector street . . . 12
2-5 Local street . . . 12
2-6 Types of traffic flow . . . 13
2-7 Service vehicles on local streets . . . 15
2-8 Subdivision with single entrance . . . 17
2-9 Subdivision with multiple entrances . . . 17
2-10 Traditional street grid . . . 18
2-11 Neotraditional street grid . . . 20
2-12 Collector street with median . . . 22
2-13 Appropriate street widths . . . 23
2-14 Too-wide street . . . 23
2-15 Street and lane widths . . . 24
2-16 Right-of-way widths . . . 26
2-17 On-street parking . . . 27
2-18 Off-street parking . . . 28
2-19 Alley . . . 29
2-20 Sight distance at crest of hill . . . 32
2-21 Horizontal curves radius . . . 33
2-22 Sight distance on a horizontal curve . . . 33
2-23 Circular turnaround . . . 34
2-24 T-shaped Turnaround . . . 34
2-25 Radius for a circular turnaround . . . 35
2-26 Off-center turnaround . . . 35
2-27 Circular turnarounds with center islands . . . 35
2-28 T- and Y-shaped turnarounds . . . 35
2-29 Auto court and eyebrow . . . 36
2-30 Shared driveways . . . 37
2-31 Streetscape treatments . . . 38

2-32 Sidewalk with grass strip 41
2-33 Path winding through a neighborhood 42
2-34 On-street bicycle lanes 43
2-35 ADA requirements for curb cuts 45
2-36 Curb cut ramps 46
2-37 Vertical curbs 47
2-38 Sloping curbs 48
2-39 Improperly designed sloping curbs 48
2-40 Asphalt curbs 49
2-41 Curbless street 49

Chapter 3: Intersections

3-1 Intersection 51
3-2 Three legged T-intersection 52
3-3 Four legged intersection 52
3-4 Realignment of angled street 53
3-5 Landing area for an acute-angle intersection 53
3-6 Traffic circles and roundabouts 54
3-7 Corner cutting 55
3-8 Intersection spacing 55
3-9 Curb radius 55
3-10 Small curb radius 56
3-11 Clear sight distance 57

Chapter 4: Streets as Drainage System

4-1 Closed drainage system 60
4-2 Open drainage systems 60
4-3 Open drainage development plan 61
4-4 Street cross-section 62
4-5 Block and lot grading 65

Chapter 5: Pavement

5-1 Street pavements 68
5-2 Asphalt streets 70
5-3 Typical asphalt street structures 71
5-4 Typical concrete street structures 71
5-5 Concrete street 72

Tables

2-1 Street Function and Average Daily Traffic (ADT) Ranges 16
2-2 Residential Trip Generation Rates 16
2-3 What Is the Cost of Excessive Street Width? 24
2-4 Recommended Pavement Widths 25
2-5 Design Speeds 30
2-6 Safe-Stopping Sight Distances 31
2-7 Minimum Rate of Vertical Curvature 32
2-8 Centerline Radius 33
2-9 Cities with Narrow Roadway Widths 50
3-1 Recommended Ranges for Curb Radii 56
3-2 Sight Distance at Intersections 57

Foreword

Attitudes toward the role of streets in a residential community have evolved from the immediate post-World War II period, when the biggest issue was how to pave dirt roads in order to bring them up to contemporary standards. The conventional practice then assumed that bigger was better. Wide streets were designed and uniform setbacks for adjacent homes established, creating an efficient street pattern for building and for moving traffic. But these practices also resulted in neighborhoods that had an unappealing sameness and where cars took precedence over pedestrians. The first edition of *Residential Streets,* developed by a partnership of the American Society of Civil Engineers (ASCE), the National Association of Home Builders (NAHB), and the Urban Land Institute (ULI), established clear principles for developers, homebuilders, and engineers based on a recognition that smaller is better in terms of cost and convenience to the resident.

Subsequent trends in planning have emphasized neighborhood design and reinforced the importance of streets as an organizing element of neighborhoods. Streets are a neighborhood issue that cannot be treated separately from the neighborhood. Neighborhood design involves the blending of several elements, including streets, with the goal of creating safe and special places in which to live. Neighborhoods should establish a sense of place for their residents. Community design involves creating closely integrated neighborhoods featuring a variety of land uses and activity centers, so that driving trips are minimized, walking is encouraged, and social and economic activities are supported. Developers are finding that homebuyers are increasingly recognizing and willing to pay a premium for superior design of the relationship among residences, streets, landscaping, and public areas. State transportation agencies, which promulgate standards for major roads that often have been applied inappropriately to the design of residential streets, now explicitly endorse the operation of local streets narrow enough to require oncoming traffic to stop and yield. This edition of *Residential Streets*—the third edition—adds a fourth partner, the Institute of Transportation Engineers, to its research and deliberations. It continues to affirm the principle that the quality of design is important in the layout and construction of streets as part of the overall planning of residential communities.

Introduction

Everyone benefits from streets that are functionally adequate, durable, and cost-effective. Builders know that inadequate or deteriorating streets can be a major cause of buyer dissatisfaction, while they also see streets as a significant element in total housing construction costs and strive to minimize their costs. Homebuyers want streets that are safe and functional yet provide an attractive residential environment. Since street costs are passed on to the homebuyer, they are an important element in the overall cost of homeownership. And the cost of maintaining the streets is an important concern to public officials and the larger community.

History

In the early 1940s, America's elected municipal officials faced public outrage over thousands of miles of alternately dusty and muddy unpaved streets that were costly to maintain. As a result, local governments adopted incentives to encourage street paving. Property owners virtually stood in line to have their streets paved on a cost-sharing basis.

After World War II, a home on an unpaved street presented a poor public image for its owner. In some areas, wide residential street pavements became prestige symbols for individuals, neighborhoods, or even entire communities. As homebuilders and land developers perceived the public's need and desire for paved streets, unpaved streets in new residential developments soon became a rarity.

FIGURE 1-1
In the 1940s and 1950s, uniform lot size and setback requirements produced repetitious street patterns and monotonous streetscapes.

Municipal administration of large street paving programs and regulation of new housing developments necessitated the adoption of street construction standards and specifications. These standards and specifications were usually patterned after readily available state highway department standards. During the immediate postwar period, formerly rural crossroads evolved into established satellite communities or large cities in their own right. The public decision makers of these new jurisdictions borrowed street-improvement standards from established neighboring cities, as minimal research into the standards-setting process had been conducted.

The standards initially adopted by cities often required the upgrading of streets to correct obvious performance deficiencies, to reduce mounting municipal street maintenance costs, or to reflect changing state highway department practices occasioned by increasingly heavy truck traffic. Nonetheless, little statistical information or research has focused subsequently on the refinement of residential street-improvement standards, although extensive studies of higher-order streets have been performed by universities, state highway departments, and the federal government. Yet residential streets represent most of the total road mileage in any region and carry a part of each trip of the vast proportion of any community's traffic.

During the 1940s and 1950s, rigid zoning ordinances and subdivision regulations that stipulated uniform lot areas and setback requirements in developing areas produced repetitious street patterns and monotonous streetscapes. In the early 1960s, changing tastes, changing values, and escalating construction costs stimulated the development and promotion of the cluster and planned unit development (PUD) housing concepts. Successful PUDs required cost-effective designs that balanced initial cost against amortization, operation, maintenance, and replacement costs. In addition, the development of PUDs dictated the satisfactory performance of street improvements, if only because of the long-term, potentially vocal involvement of organized PUD residents. Further, cluster planning and PUDs drew the public's attention to the values associated with open spaces and the landscape, including the streetscape. This early focus on the landscape has had a significant influence on today's increased environmental awareness.

By the early 1960s, improved land planning for suburban areas had attracted serious attention. Many professional and industry groups, government agencies, and universities studied and evaluated innovative community and neighborhood design. The planned unit development concept was refined by planners and developers and today is widely accepted by consumers and local governments.

One factor in the acceptance and success of cluster housing and PUDs was the latitude accorded planners and engineers to lay out and interrelate street patterns, open spaces, and building sites. The design flexibility inherent in PUD planning has made possible the preservation of natural vegetation and topography and has permitted sound economies in land development, with benefits accruing to both individual residents and the community at large.

Early Standard Setting

Public officials and professional associations generally focused on guidelines that are most reasonable for major thoroughfares, but are excessive for local residential streets. The traffic characteristics of residential streets and their construction, maintenance, and performance requirements differ significantly from arterial and collector roads.

Between 1936 and 1941, in order to upgrade local residential street paving practices, the Federal Housing Administration (FHA) set forth street improvement requirements for FHA-approved land developments in all metropolitan areas and many nonmetropolitan communities. The guidelines established either FHA standards or local community standards, whichever were more stringent, as the minimum acceptable practice. The FHA discontinued application of these standards in 1967, as the problem of "no local standards" had been supplanted by another problem: excessively rigorous local standards that added needlessly to costs.

In light of FHA's fundamental objective to improve housing conditions and the quality of the living environment, it is significant that an important motivation behind the agency's abandonment of its street improvement requirements was its unwillingness to endorse economically unjustifiable construction requirements. FHA's decision must be viewed in perspective, recognizing that, by 1967, even small communities required that all newly platted residential streets be paved and had adopted pavement design and quality standards.

Because the homebuilding industry acknowledged the value of streets in the presentation of the housing product, it supported FHA's early street improvement objectives. Streets are a particularly visible element within the residential setting and their design and appearance can greatly enhance or detract from a subdivision.

More Recent Efforts

The 1970s and 1980s saw efforts to develop new residential street design guidelines that, unlike previous standards, were not based upon highway standards. In

FIGURE 1-2
Street design as art. An arterial road in Anaheim, California.

MICHAEL A. BAUSER

FIGURE 1-3
A street scene in Ferrara, Italy.

1974, the American Society of Civil Engineers (ASCE), the National Association of Home Builders (NAHB), and ULI (the Urban Land Institute) published *Residential Streets: Objectives, Principles, and Design Considerations,* the first book in a series that this book continues. Though not a comprehensive design guide, the 1974 book included a discussion of significant factors that should be considered in the design of residential streets. It focused on the relationship between the design of residential streets and their unique function, the cost-effectiveness of street design, and the role of streets in establishing a residential community's intimate scale rather than detracting from it.

Performance Streets, a book published by the Bucks County (Pennsylvania) Planning Commission in 1980, represents a significant and still valid contribution to the body of work on residential street design. The book reviewed and evaluated the street design literature and standards developed by other authors and organizations, offered its own recommendations, and suggested ordinance language for use by local governments in implementing the recommendations. The book was based on the concept that the movement of vehicles is only one of a residential street's many functions. A residential street is also part of its neighborhood and provides a visual setting for the homes as well as a meeting place for residents.

In 1984, the Institute of Transportation Engineers (ITE) published *Recommended Guidelines for Subdivision Streets,* which stressed four factors—safety, efficiency of service, livability, and economy—as guides in the design of residential streets. Among the many principles that ITE derived from consideration of these factors, two are particularly notable: 1) local streets should be designed to discourage excessive speeds; and 2) the land area devoted to streets should be minimized. The ITE guidelines were updated in 1990 (*Proposed Revisions to*

FIGURE 1-4
West Orange Trail, Orange County, Florida.

Guidelines for Residential Street Design) and again in 1993 (*Guidelines for Residential Street Design, a Recommended Practice*).

In 1990, the American Association of State Highway and Transportation Officials (AASHTO) published *A Policy on Geometric Design of Highways and Streets.* (A metric version was published in 1994 and a new edition was in preparation as of early 2001.) Commonly called the AASHTO Greenbook, this manual is the definitive source of guidelines for local road design—and not only, as is often thought, for major collector roads and arterials. The AASHTO Greenbook addresses local streets in considerable detail in its "Local Urban Streets" section. The single most important feature of residential street design—the ability for traffic moving in opposite directions to share a single traveling lane—is clearly articulated in the AASHTO Greenbook.

During the 1980s, many of these evolving concepts about the proper role of streets in a neighborhood and the appropriate design of local streets were applied in new residential developments. In particular, the Joint Venture for Affordable Housing, a program initiated by the U.S. Department of Housing and Urban Development (HUD) and supported by many other organizations including NAHB and ULI, advocated the application of a number of the principles that had evolved during the 1980s as part of an overall strategy to reduce housing costs.

Philosophy of Residential Street Design

In the past, residential streets have been mistakenly viewed as fulfilling only three functions: providing access, providing on-street parking, and conveying traffic. As a consequence, requirements and design guidelines have placed undue stress on the efficient movement of traffic (on moving traffic in greater volumes or at increased speed) and have ignored the many other functions of residential streets. As stated in *Performance Streets* (Bucks County, 1980):

It was often forgotten that residential streets become part of the neighborhood and are eventually used for a variety of purposes for which they were not designed. Residential streets provide direct auto access for the occupant to his home; they carry traffic past his home; they provide a visual setting, an entryway for each house; a pedestrian circulation system, a meeting place for the residents, a play area (whether one likes it or not) for the children, etc. To design and engineer residential streets solely for the convenience of easy automobile movement overlooks the many overlapping uses of a residential street.

Residential Streets, Third Edition, is based on the premise that the design of a residential street should be appropriate to its functions. A residential street's functions include not only its place in the transportation system but also its role as part of a residential community's living environment.

The idea of a residential street system as much more than a transportation facility is reflected in the following principles that form the basis for the guidelines presented in this book:

- Street planning should relate to overall community planning, including pedestrian and bicycle activity.
- Traffic in residential areas should be kept to a minimum to reduce noise, congestion, and hazards to pedestrians.
- The street is an important component of overall residential community design. Properly scaled and designed streets can create more attractive communities and can contribute to a clearly defined sense of place.
- Street design standards should permit flexibility in community design. They should allow street alignments to follow natural contours and preserve natural features or to respond to other design objectives such as the creation of intimate urban- or village-scale streetscapes.
- Wherever possible, street pavement layouts should be planned to avoid excessive stormwater runoff and to avoid heat buildup.
- Streets can function socially as meeting places and centers of community activity. For example, children often use low-volume-traffic streets as play areas.
- In the interest of keeping housing affordable, street costs should be minimized.
- The overdesign of streets should be avoided. Excessive widths or an undue concern with geometry more appropriate for highways encourages greater vehicle speeds.
- Different streets have different functions and need to be designed accordingly. Blanket guidelines are inappropriate.

These principles suggest that a street system should be designed as a hierarchy of street uses. Routes carrying through traffic should be separated from routes that provide access to residential properties. All streets can be described in terms of what kind of traffic they serve at what level, which is known as their functional classification. Interstate highways, for example, are designed for high-speed through traffic but provide no direct access to properties. At the other end of the functional classification spectrum are small residential streets designed only for access. These streets should be considered part of the residential neighborhood rather than part of the traffic system. In residential street design, the street's contribution to the neighborhood environment is as important as the street's role as a transportation link. In addition, the street system should be designed to be easily read (or understood) by users so that the intended function of a particular street segment is readily apparent.

Public and Private Streets

Most residential streets are public streets. Usually, developers dedicate them to the local government upon completion or, sometimes, after a prescribed period of time—often one year—has passed. The local government is then responsible for maintaining the streets. In some developments, however, a homeowners association or community association owns and maintains some or all of the streets as private streets.

The ownership of the street should not be a factor in its design or function. The width of a street, for example, should be based upon the volume and characteristics of the expected traffic and on the likely amount of on-street parking. Some communities permit flexibility in the design of private streets but have not changed their guidelines to allow the appropriate design of public streets. The design recommendations in this book are meant to apply to all residential streets regardless of ownership and maintenance responsibility.

Applications

The design principles outlined in this book are applied most easily to new communities, especially to planned communities in which a single developer designs the entire circulation system and community plan to complement and support one another.

However, the same design principles can be applied also to new developments under multiple ownership. Under local government guidance and control, the coordinated development of numerous sites can create a unified community pattern that offers advantages similar to those enjoyed by communities planned by a single developer. Many of these design principles can be used also to redesign traffic patterns in older neighborhoods to improve traffic safety and enhance neighborhood livability.

To apply the design concepts in this book, communities may need to rethink the proper role of residential streets and change some existing guidelines, regulations, and policies. The potential result—safe and cost-effective streets that help create more attractive and livable neighborhoods—may be worth the effort required.

2

Design Considerations

Typically, the design process begins with the layout of the street system—identification of the general location of the street corridors and their functional classification. This layout is established primarily through reference to the natural features of the site, the site's connection points to the external road system, and a preliminary idea of the lot plan.

The next step in the design process is the identification of the cross-section features—the number of lanes, the type of drainage (curb/gutter, open swale, other), the pavement width, the sidewalk placement, and the right-of-way needed to contain the street elements. The design of the cross-section features is driven largely by the functional classification selected for the individual streets.

Finally, the longitudinal features—the horizontal alignment, the vertical profile, and the gradient—are designed. These features are largely controlled by the details of the site, primarily the topography and the detailed lotting plan.

Functional Classification System for Streets

A fundamental concept in the geometric design of a highway or street is its use or functional classification. Many design professionals classify roadways in terms of the balance they offer between mobility (measured by relatively long-distance travel at relatively high speeds) and access (measured by service to origins and destinations). The commonly used hierarchy of streets includes, in ascending order, local streets, collectors, and arterials (including all freeways).

FIGURE 2-1
Hierarchy of streets.

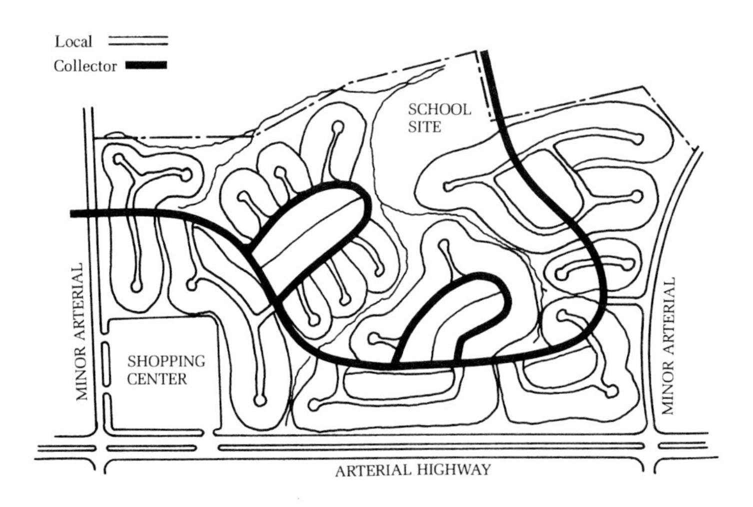

Residential frontage along streets may occur in any of these categories except freeways (Figure 2-1). A functional description of each class of streets follows:

Freeways. Freeways are intended primarily to serve through traffic traveling relatively long distances (Figure 2-2). They provide no access to adjacent land except at interchanges, which are spaced fairly widely.

Arterial Streets. The primary purpose of arterial streets is mobility—the movement of as much traffic as possible as fast as is reasonable (Figure 2-3). The mobility function of arterials overshadows their function of providing access to fronting properties. Paradoxically, the high volume of traffic on arterial streets is an inevitable invitation to commercial land uses to locate along these streets. Arterial street mileage is almost never created as a part of new subdivisions. Consequently, for several decades, new arterial streets fronted by residen-

FIGURE 2-2
Freeways do not provide access to land, but they serve local trips going relatively long distances.

BOB DUNPHY

SUMMERLIN

FIGURE 2-3
On arterial streets, the mobility function overshadows the function of providing access to adjacent properties.

tial properties have been rare. Recent development activities related to the new urbanism trend—for example, the revival of luxury apartments in urban settings, the conversion of nonresidential buildings into residential lofts and other residential uses, and the success of in-town traditional neighborhood developments (TNDs) in some cities—have reintroduced the concept of designing residences to front on arterial streets. The level of such activity is small but attention-getting. The design of arterial streets is not addressed in this book. For material on the design of arterial streets, see Chapter 7, "Policy on the Design of Arterial Streets and Highways," in the AASHTO (American Association of State Highway and Transportation Officials) Greenbook .

Residential Collector Streets. The function of residential collector streets is a balance between mobility and access (Figure 2-4). Residential collector streets typically serve as the link between local streets and arterial streets. A typical vehicle trip is likely to include a short segment of travel on residential collector streets.

Traffic volumes on residential collector streets usually exceed 1,500 vehicles daily, although traffic volume is not the prime determinant of a collector street. Increasingly, new collector streets are fronted by active properties—typically businesses, institutions, or multifamily residences. Properly designed, residential collector streets can have high value as frontage for both residential uses and small-scale neighborhood commercial uses.

Typically, residential collector streets make up about 5 to 10 percent of the total street mileage in new development. Ideally, no residential location is more than one-half mile from a collector street. The ability to reach community travel destinations—daily shopping, daycare, elementary schools, and so forth—using

FIGURE 2-4
The entrance boulevard serves as a collector street for South Riding in Loudoun County, Virginia.

TORTI GALLIS. COURTESY OF SOUTH RIDING

no streets higher on the hierarchy than residential collector streets is a valuable and highly salable feature of residential communities. Much of the appeal of travel in small towns is due to their abundance of residential collector streets. New communities that successfully mirror the driving atmosphere of small towns invariably do so by linking homes with destinations largely through the use of residential collector streets.

Local Streets. Local streets are provided predominantly for access to the residential properties that front them (Figure 2-5). Mobility—the ability to travel relatively long distances at relatively high speeds—is not a priority on local streets. Local streets usually account for around 90 percent of the street mileage in new communities. However, they generally account for only a small fraction of total vehicle-miles traveled. As the preponderant class of street in terms of mileage in residential areas, local streets contribute much to the physical signature of their neighborhoods. They also constitute the backbone of neighborhood pedestrian and bicycle networks.

FIGURE 2-5
Local streets provide access to adjacent properties. Providing mobility is less important.

HNTB

This book addresses the design of residential collector and local streets. Streets intended to serve other land uses, such as major retail, office, or industrial complexes, differ dramatically in design from residential streets. The definitive source of design guidelines for streets serving other land uses is AASHTO's *A Policy on Geometric Design of Highways and Streets* (1990; the AASHTO Greenbook).

Types of Traffic Flow

The street types recommended in this book will produce one of the following three types of traffic flow (Figure 2-6):

Free Flow. Free-flow traffic requires providing each direction of moving traffic with its own designated lane, a lane that is not shared with parking or traffic moving in an opposite direction. Drivers moving in one direction need not interact with drivers going in the opposite direction. A free-flow street typically has a marked center line. The traffic on residential collector streets is typically free flow.

Slow Flow. Slow-flow operation occurs when parked vehicles or other conditions constrain the space for vehicles moving in opposite directions to pass. Drivers react to motorists coming from the opposite direction. Some drivers may choose not to pass an oncoming vehicle, but rather to stop and yield the right-of-way. Typically, traffic engineers do not mark a center line, except perhaps at

FIGURE 2-6
Types of traffic flow.

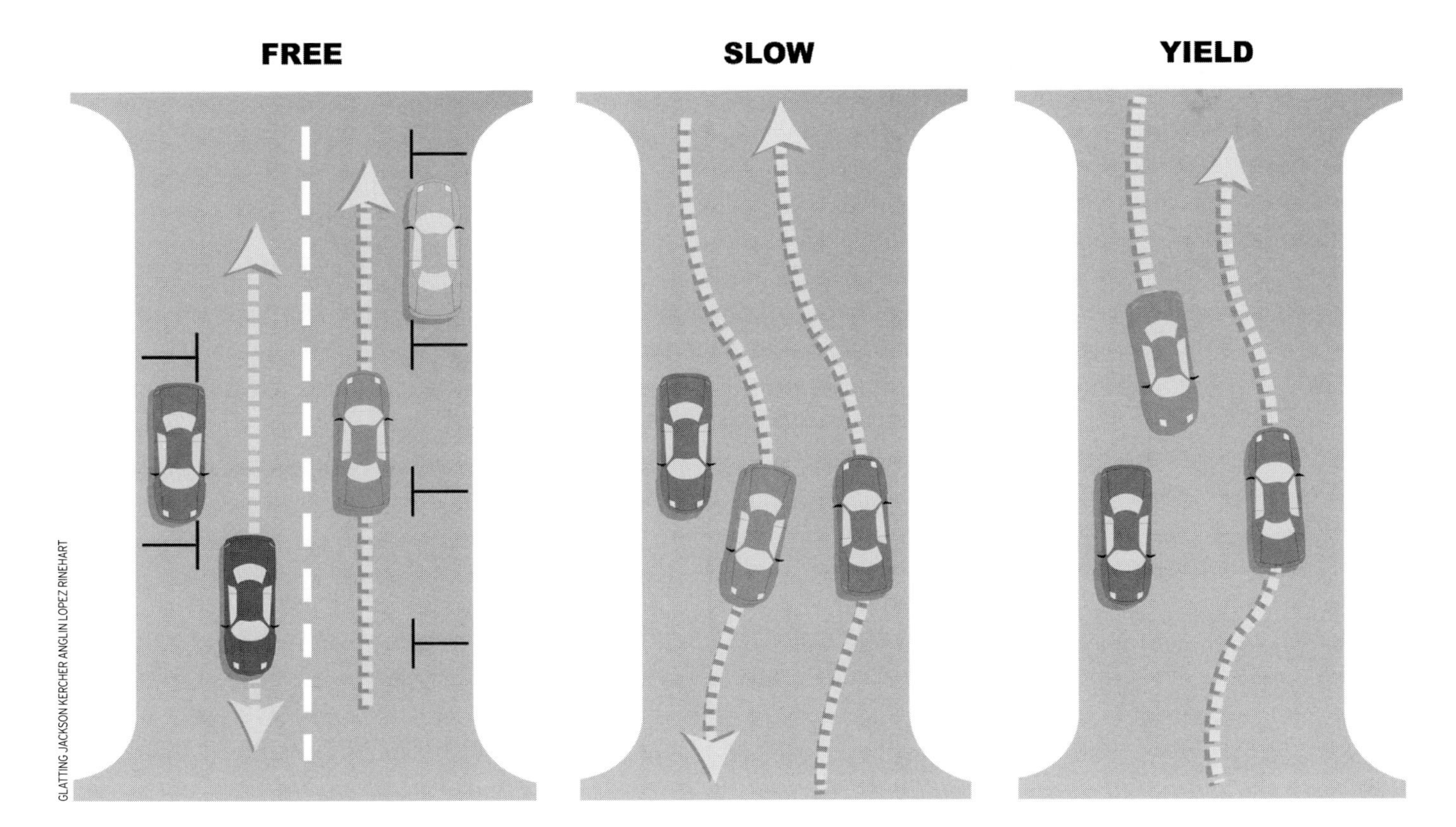

intersection approaches, thus acknowledging that motorists traveling in opposite directions can share the same pavement. Local streets with high ADT (average daily traffic, or the total number of vehicles traveling in both directions past a point on the street on a typical day) and some residential collector streets may have slow-flow operation.

Yield Flow. Yield flow occurs when two-way traffic is impossible where parked vehicles are present. Thus, some motorists must stop and yield the right-of-way to oncoming vehicles. For decades prior to the 1960s, yield flow was the widely accepted norm for local streets. Thus, the vast majority of local streets throughout the United States operate at yield flow. Yield operation continues to be explicitly endorsed in the AASHTO Greenbook, which acknowledges the condition of "one unobstructed moving lane [where] opposing conflicting traffic will yield and pause on the parking lane area until there is sufficient width to pass." Most local streets with low ADT may have yield-flow operation.

Emergency Vehicle Access

The street design guidelines in this book fully accommodate the emergency vehicles (fire, rescue, emergency medical, and towing) that serve residential areas.

For emergency vehicles, the most important element of street design is the accessibility afforded by a well-connected network of streets. The street pattern should provide at least two routes to any street within the community, reducing the risk that an emergency vehicle will find its access blocked. A street pattern composed of a single entry/exit street spine with all destinations located on dead-end branches of the spine presents difficult challenges for emergency vehicle access.

All of the streets recommended in this book provide adequate pavement width for emergency vehicles. The most confining street situation for emergency vehicles is the local street with cars parked on both sides. The parked cars occupy 13 to 14 feet of the roadway, leaving ten to 13 feet for the passage of emergency vehicles, even on a minimal 24- to 26-foot-wide street. The maximum width of a standard pumper is eight feet, excluding mirrors. Thus, even with parked vehicles present on both sides of a local street, a standard pumper can freely negotiate the street (Figure 2-7).

The published guidelines from professional sanctioning bodies, such as the National Fire Protection Association, do not specify pavement widths or any other dimensions for roadways, but rather regard guidelines for street dimensions as a local government prerogative. The AASHTO Greenbook addresses fire equipment, categorizing residential pumpers as single unit (SU) vehicles. SU vehicles are fully accommodated in all of the guidelines recommended in *Residential Streets,* Third Edition.

Should access for emergency or maintenance vehicles become a problem on small residential streets, a variety of operational measures (in lieu of wider

WALKABLE COMMUNITIES INC.

GLATTING JACKSON KERCHER ANGLIN LOPEZ RINEHART

FIGURE 2-7
Standard pumper (at left) and trash collection (at right) on local streets.

streets) can provide relief. Establishing no-parking areas on some of a narrow street's frontage can open up space for vehicles to pass and for setting up emergency equipment. Parking can be eliminated on one side of streets that are frequently congested. Alternate-side parking restrictions, which often are used to facilitate street cleaning and snow removal, can be applied to new as well as existing streets.

The proper sizing of emergency equipment can do much to meet the emergency vehicle access challenge. A one-size-fits-all philosophy of specifying fire equipment may put needlessly large vehicles in residential use. A more productive approach than redesigning streets to fit the vehicles would be to use vehicles that fit the community's needs. For example, tower trucks feature outriggers that require clear spaces of up to 20 feet wide, making them inappropriate vehicles for residential use. Rather than accommodating outriggers with a street design that detracts from the livability of neighborhoods, the fire equipment itself should be designed to be appropriate for the structures in the neighborhood and thus to accommodate the neighborhood.

Many large vehicles regularly visit residential streets—for solid waste removal, recycling pickups, school transportation, parcel delivery, snow removal, and utilities work. Automated solid waste collectors and many other such vehicles have evolved technically to enlarge their payloads without making increasing claims on street width. It is reasonable to expect emergency equipment to similarly adapt its design to the street widths that are encountered.

Traffic Volumes and Residential Dwelling Units

In properly designed residential neighborhoods, travel distances from residences to collector streets should be short, traffic speeds should be low, lane capacity and design speed should not be controlling design factors, and inconvenience or minor delay should be acceptable. Further, drivers and residents expect

TABLE 2-1

STREET FUNCTION AND AVERAGE DAILY TRAFFIC (ADT) RANGES

	ADT Range	Dwellings Served[1]
Local Streets	400–1,500	40–150
Residential Collector	>1,500	>150

[1]Based on single-family detached houses, at ten daily trips per dwelling unit.

brief delays and accept the need to decrease speed. In fact, it is customary for responsible individuals to drive carefully to avoid children and pets.

Average daily traffic (ADT), the total number of vehicles traveling in both directions past a point on a typical day, can help guide the choice of street type. The number of dwelling units served by the street (that is, using it as the preferred route) is another factor that can help guide the choice of streets. Usual ADT and number of dwelling units served by different classes of streets are presented in Table 2-1. The ADT range and housing units served for different classes of streets may overlap, and thus are not intended to serve as absolute design criteria.

The traffic density and speed found on highways, arterials, and collector streets are absent from local streets, and driving attitudes and habits on local streets differ from driving behaviors on highways, arterials, and collector streets. Yielding momentarily to resolve minor traffic conflicts is practical at the speeds observed in residential areas. In residential areas, traffic yields to drivers backing from their driveways or drivers coming out of their driveways yield to oncoming traffic, and no one is unduly delayed. If parked vehicles impede residential traffic, approaching vehicles often yield and then proceed with caution. Street design that encourages this kind of cautious driver behavior can result in reduced speeds and more attentive drivers, and thus make streets safer.

The primary considerations in selecting guidelines for residential streets, therefore, are the characteristics of local residential traffic and the expectations of residents. Traffic volumes can provide additional guidance for decision making.

TABLE 2-2

RESIDENTIAL TRIP GENERATION RATES

	Vehicle Trips per Dwelling Unit	
	Weekday	Peak Hour
Detached Single-Family Units	9.6	1.00
Apartment Units		
• All Apartments	6.6	0.67
• Low-Rise Apartments	6.6	0.62
• High-Rise Apartments	4.2	0.40
Townhouse and Condominium Units	5.9	0.54

Source: Institute of Traffic Engineers, *Trip Generation Handbook*, Sixth Edition (Washington, D.C.: ITE, 1997).

The Institute of Traffic Engineers, the definitive source of trip generation data, reports rates of trip generation for types of residential units as shown in Table 2-2.

Neighborhood Access to Adjacent Street System

Entrances to residential areas from arterial or collector streets should be designed to allow convenient access without encouraging through traffic, and they should provide for safety and convenient turning. If these entrance streets allow access to community facilities, retail areas, and residential streets, they may require left-turn lanes located at major access or discharge points.

Opinion differs as to whether residential areas should have several entrances from arterial or collector streets (Figures 2-8 and 2-9). Multiple access points make possible alternative travel routes and thus offer several advantages: reduced congestion and internal travel volumes; diffusion of the development's traffic impact on the external road system; and continuity in the internal street system for emergency and delivery services and for snowplows and other maintenance vehicles.

FIGURE 2-8
Single-entrance subdivision (at left).

FIGURE 2-9
Subdivision with multiple entrances (at right).

DAVID JENSEN ASSOCIATES, INC.

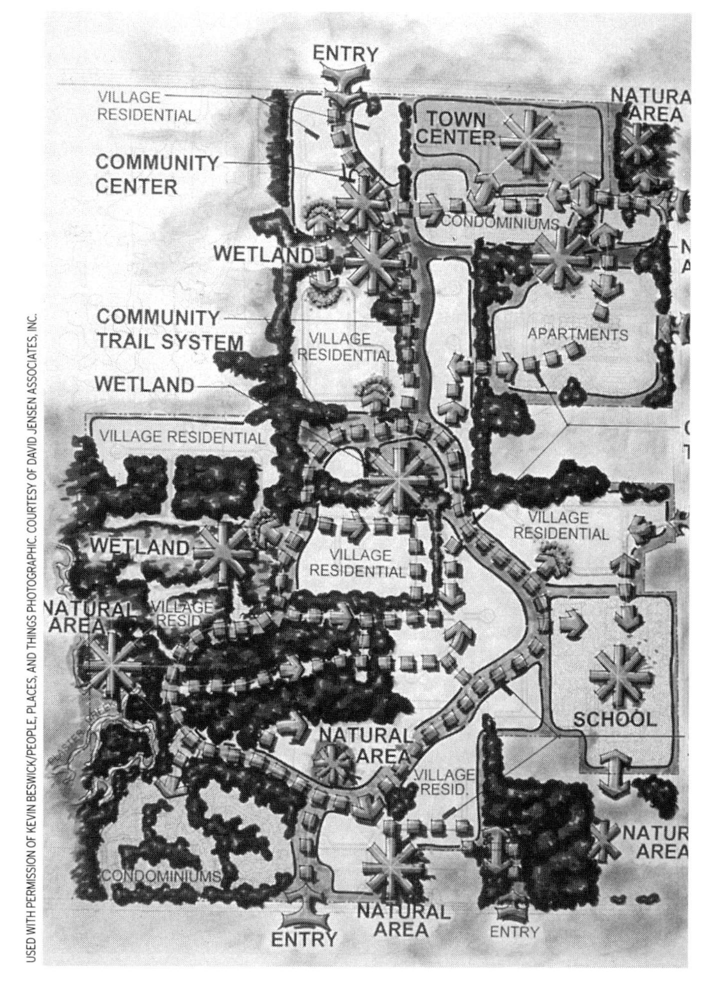

USED WITH PERMISSION OF KEVIN BESWICK/PEOPLE, PLACES, AND THINGS PHOTOGRAPHIC. COURTESY OF DAVID JENSEN ASSOCIATES, INC.

The advantages of having a single access point include the elimination of through traffic and short-cutters, increased security, and a greater sense of community identity. However, as development density increases, community street systems that focus all of the community's traffic on a limited segment of the connecting street system become unworkable.

Street Layout and Platting

Decisions regarding street layout should result from evaluations of several factors, including topography, soil and geologic conditions, drainage patterns and potential runoff quantities, length and type of streets, purpose of individual streets, and desired design character. Residential development should present neither an endless vista of trafficways that encourage through traffic nor a spaghetti-like labyrinth that is irrational, incomprehensible, and confusing.

Adequate topographic mapping is a fundamental requirement for optimum residential neighborhood planning, including street layout. The planner should be sensitive to development, construction, operation, and maintenance costs that often can be minimized by properly interrelating the street layout with the natural topography. Basic topographic maps, overlaid with supplemental information on soils, depths to rock, groundwater conditions, specimen-tree locations, desirable natural features, peripheral occupancy characteristics, and so forth, provide the land planner with integrated information upon which to base designs that are functionally, economically, aesthetically, and environmentally optimal.

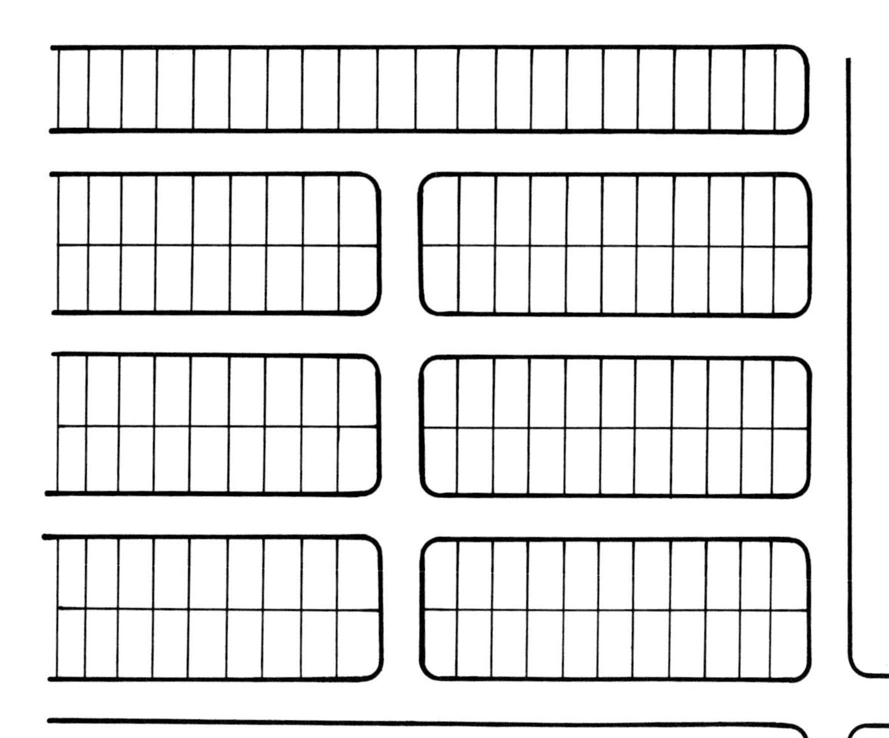

FIGURE 2-10
Traditional grid with long, straight streets.

In the housing boom that followed World War II, most new suburban development was designed along grids of straight streets, following the patterns of traditional towns and cities (Figure 2-10). This grid pattern of streets developed as cities grew rather than as a result of regulation, and it proved to be well adapted to several decades of increasing automobile use.

However, the great majority of residential subdivisions developed since the 1960s have been laid out in a pattern that is fundamentally different from the traditional city grid. Instead of a dense network of fully connected streets, these subdivisions adopted spare street layouts, with mostly dead-end streets and culs-de-sac and only one or two entry points. Street alignments were generally curved, either to adapt to the terrain or simply to present a more rural appearance. Throughout the 1960s and 1970s, this type of street pattern was thought to be ideal. It provided privacy for homeowners on the dead-end streets, minimized the amount of paving that the developer needed to do, eliminated all but local traffic on many streets, and, at least at first, seemed to preserve a rural environment.

However, the benefits associated with the postwar style of suburban development have been seen to be eroding since the start of the 1980s. Because of the lack of internal street connections, all traffic—even for the most local of destinations—is funneled out onto the surrounding arterial street system. Consequently, the surrounding street system becomes congested even at low levels of development, undermining the small-town or rural atmosphere sought by developers and homebuyers. Funneling all traffic onto a limited amount of arterial highways creates an almost unstoppable pressure for locating all commercial development in a strip along these highways, which further erodes the suburban landscape and exacerbates traffic problems on the arterial highways. The quality of life of subdivision homeowners is greatly diminished by having to travel on congested arterial highways for all of their daily travel needs. The lack of direct routes between destinations that is a feature of unconnected street systems is a major obstacle to pedestrian and bicycle travel. Also, unconnected streets create inefficiencies for school bus travel, transit services, and municipal services of all types. By the 1990s, widespread concern that the 1960s type of residential subdivision was proving to be unsustainable led people to conclude that the model of the postwar subdivision should be rethought.

In response, a trend toward what was called neotraditional development or traditional neighborhood development emerged in the 1980s and 1990s (Figure 2-11). Street patterns in neotraditional community design capture much of the connectivity of older city grids, but with few of the long, straight, continuous streets of those grids. Neotraditional community designers seek to create a fabric of local streets so that much of the community's daily driving can be accommodated without the use of the surrounding arterial street system. The highly

FIGURE 2-11
Kentlands, in Gaithersburg, Maryland, is a neotraditional development that incorporates a traditional grid pattern of streets.

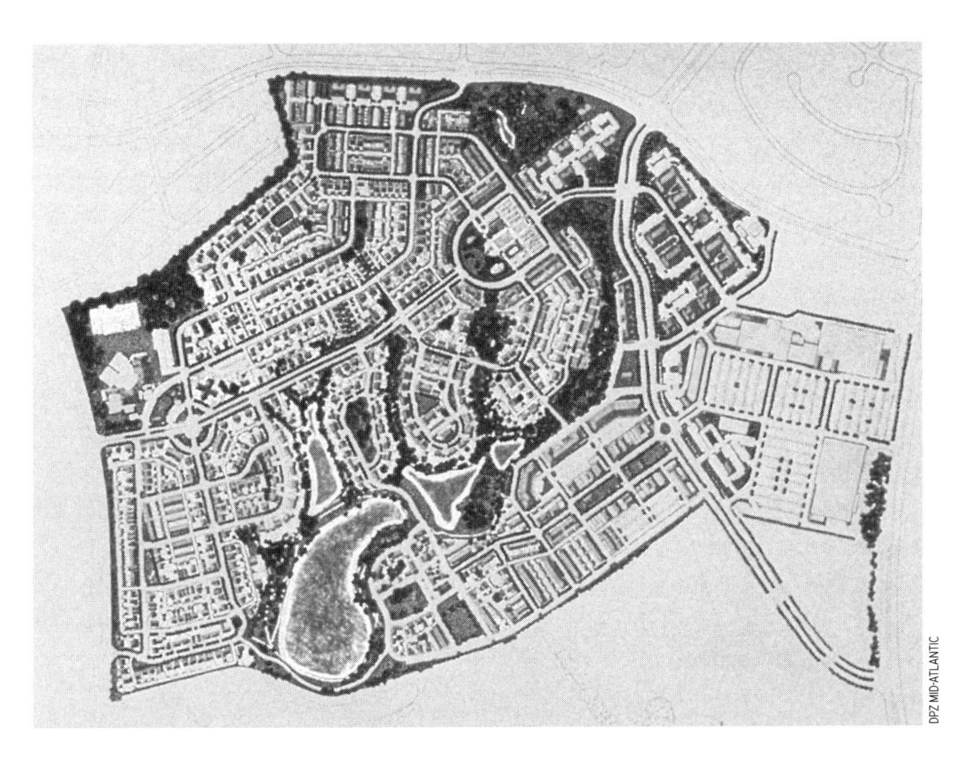
DPZ MID-ATLANTIC

connected street network in neotraditional developments and the small, low-speed character of the streets within that network give bicycle and pedestrian travel a high level of service.

The growing experience with neotraditional development has engendered some different approaches to connectivity. Connecting the entire network of local streets from one subdivision to the next, even where mandated by local regulations, has proved to be a near impossibility, politically. Options for establishing connectivity while preserving neighborhood identity and privacy include creating a framework of residential collector streets on public rights-of-way. With neighborhood connectivity thus assured, the individual residential communities can arrange their internal streets in a variety of ways, needing only to meet the residential collector network at limited points.

Principles of Street Layout

As stated in *Guidelines for Residential Subdivision Street Design* (Institute of Transportation Engineers, 1993), residential street layout is guided by a few basic principles:

- safety for both vehicular and pedestrian traffic;
- efficiency of service for all users;
- livability of the residential environment; and
- economy of land use, construction, and maintenance.

The street layout is an integral part of the success of a residential community in terms of both function and marketability of the houses built. The street layout determines to a great extent the location of utility installations, the solar orientation of the houses, the degree of interaction among neighbors, and many other community features. ITE's principles for the layout of residential streets translate into the following design guidelines.

- Paved access should be provided to all developed parcels.
- Through traffic, defined as traffic with neither origin nor destination in the neighborhood, should be discouraged.
- The creation of excessive travel lengths should be avoided.
- Local street systems should be logical and understandable.
- The community's circulation system and land development patterns should not cause unnecessarily large volumes of local traffic to be routed onto adjacent major streets (thereby detracting from their efficiency).
- The local circulation system should not have to rely on extensive traffic regulations or control devices to function efficiently and safely.
- Traffic generators—such as schools, churches, and neighborhood shops—within residential areas should be considered in the local circulation pattern.
- Residential streets should clearly communicate their local function and place in the street hierarchy.
- The local street system should be designed for a relatively uniform low volume of traffic.
- To discourage excessive speeds, streets should be designed with curves, changes in alignment, and short lengths. And streets should be designed to be no wider than is necessary.
- A minimum amount of space should be devoted to streets.
- The number of intersections should be limited.
- Local street layout should permit the economical development of land and the efficient layout of lots.
- Local streets should be responsive to topography and other natural features.
- Public transit service should be provided where appropriate in residential areas.

Number of Lanes

The design for local streets should ensure at least one unobstructed moving lane, shared by vehicles moving in either direction, even if parking is present on both sides. The level of user inconvenience occasioned by the lack of two moving lanes (that is, by slow- or yield-flow operation) is remarkably low in areas where single-family units predominate. Oncoming traffic pauses and yields in the parking lanes until there is sufficient room to pass. Local residential street patterns are such that travel distances are usually less than one-half mile between the point of trip origin and a free-flowing residential collector street. In high-density multifamily residential areas, two moving lanes (that is, free-flow operation) may be required to accommodate opposing traffic.

FIGURE 2-12
Residential streets can be divided with a median to provide an attractive entryway.

A major distinction between local streets and residential collector streets revolves around parking considerations. Local streets allow parking even if doing so reduces the street to a single moving lane. The parking on residential collector streets, on the other hand, must be located in dedicated parking lots, leaving two unobstructed moving lanes at all times to accommodate greater traffic volumes.

Residential streets can be divided with a median to preserve a desirable natural feature, minimize the necessity of grading on steep terrain, or provide an attractive entryway (Figure 2-12). When medians are used, planners must design for nighttime visibility, particularly at the ends of the median. Paired one-way residential streets, even single-frontage streets, may be necessary in areas characterized by steep terrain.

Pavement Widths

Residential street designers should select the minimum width that will reasonably satisfy all realistic needs, thereby minimizing construction and annual maintenance costs, while at the same time maximizing the livability of the community (Figure 2-13 and Table 2-3). The tendency of many communities to equate wider streets with better streets and to design traffic and parking lanes for free-flow traffic is a highly questionable practice. Certainly providing for the free flow of traffic in two 11- or 12-foot lanes that are never occupied by parking can encourage traffic to speed (Figure 2-14). Encouraging slower traffic speeds through narrower streets can improve the safety of streets for residents. Some studies indicate that as a street becomes wider, accidents per mile increase exponentially; and that the safest residential street may be a narrow street.

FIGURE 2-13
Street widths should be consistent with traffic needs.

On most local streets, a 24- to 26-foot-wide pavement is the most appropriate width. This provides either two parking lanes and a traffic lane (yield-flow operation) or one parking lane and two moving lanes (slow-flow operation). For lower-volume streets with limited parking, a 22- to 24-foot-wide pavement is adequate.

For low-volume local streets where no parking is expected (for example, large-lot, rural communities), an 18-foot pavement is adequate. Widening access streets a few more feet does not significantly increase capacity, but it does permit wider moving lanes that tend to encourage higher driving speeds. A wide access street also lacks the intimate scale that makes an attractive setting for housing.

FIGURE 2-14
Streets that are too wide are unattractive and encourage unsafe speeds.

TABLE 2-3

WHAT IS THE COST OF EXCESSIVE STREET WIDTH?

Not only do excessive street widths reduce the livability of a community, but they also entail additional costs that must be paid by homeowners. The figures cited here are for 2001, based on unit costs of contractor services for a project in northern California. For this project, a section of street 100 feet long would cost about $9,500 to build to a width of 24 feet, compared with $13,500 for 36-foot-wide street. Paving widths are 20 feet and 32 feet, respectively, with an additional two-foot gutter on each side. Moreover, in this area where lots sell for $300,000 per acre, the land costs actually exceed the street construction costs, even for the narrower street. Land and construction total costs for a 100-foot section of a 36-foot-wide street amount to almost $40,000, compared with $26,000 for a narrower 24-foot street.

	Cost per 100 Feet of Street	
	24′ Wide	**36′ Wide**
5-Inch Asphalt Paving/6-Inch Base	$6,800	$10,880
6-Inch Curb and Gutter	1,265	1,265
4-Inch Sidewalk	1,400	1,400
Total Construction Costs	9,465	13,545
Land (at $300,000/acre)	16,800	25,200
Total Cost	$26,265	$38,745

FIGURE 2-15
Street and lane widths.

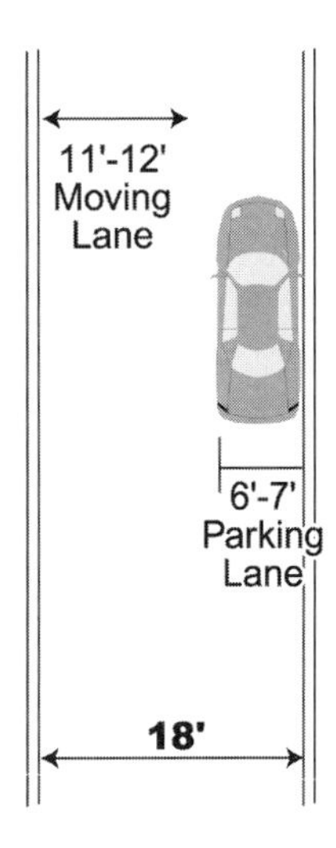

Local
(Parking not expected or restricted to one side)

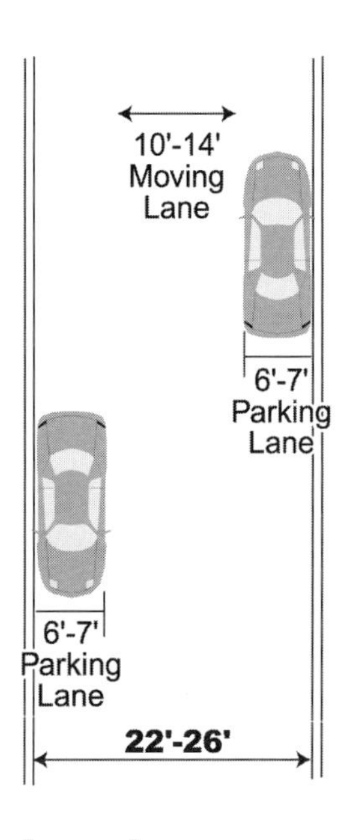

Local
(Parking on both sides)

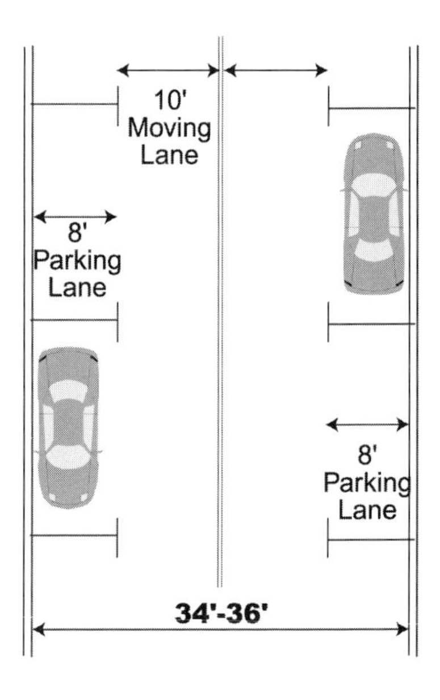

Residential Collector
(Parking on both sides)

GLATTING JACKSON KERCHER ANGLIN LOPEZ RINEHART

TABLE 2-4

RECOMMENDED PAVEMENT WIDTHS

	Pavement Width
Local Streets	
• No Parking Expected	18 feet
• Low or Restricted Parking	22–24 feet
• Normal Residential Parking	24–26 feet
Residential Collector	32–36 feet

Once traffic from tributary local streets has reached a volume needing two clear traffic lanes, a street becomes a residential collector. A residential collector street should be designed for higher speed than local or access streets, permitting unrestricted automobile movements. Residential collector streets with a pavement width of 36 feet provide for adequate traffic movement and two curb parking lanes (Figure 2-15). Where houses do not front on the residential collector street and parking is not normally needed, two moving lanes of pavement are adequate, with shoulders graded for emergency parking. Table 2-4 summarizes these pavement width recommendations.

Right-of-Way Widths

The right-of-way should be only as wide as necessary for the street pavement and other facilities and uses, including sidewalks, utilities, drainage, street trees, snow storage, and grading (Figure 2-16). The width of rights-of-way is often mandated by ordinances, and requirements should reflect the real needs of the selected design.

A right-of-way width allowance for future street widening is unnecessary in well-planned residential neighborhoods. Since true residential streets are not subject to widening, the prospect of future widening is not a factor in specifying right-of-way dimensions. If a street is subject to future widening by virtue of its location and traffic routing, it should not be functionally classified or treated as a residential street.

When the public right-of-way includes sidewalks or bicycle paths, it needs to be wide enough to accommodate these uses. If additional right-of-way is desired for utility maintenance, an additional one foot outside these improvements is sufficient. Another option is to locate pedestrian walks and bicycle paths on land owned in common by the community association or on easements on private property rather than in the right-of-way.

Many jurisdictions require the removal of all trees within a very wide right-of-way. In developments built on wooded land, such a practice results in a great loss of trees and increased expense in clearing. Limiting the width of the right-of-way to only what is necessary for the safety of the traveling public and clearing trees only within three to five feet of the pavement edge are design practices that can yield substantial benefits by helping to create a more attractive community. The AASHTO Greenbook guidelines permit trees to within 1.5 feet of the back of the curb.

On-Street Parking

There are three options for on-street parking: 1) parking on one side of the street, 2) parking on both sides of the street, and 3) parking allowed only in

FIGURE 2-16
Right-of-way widths.

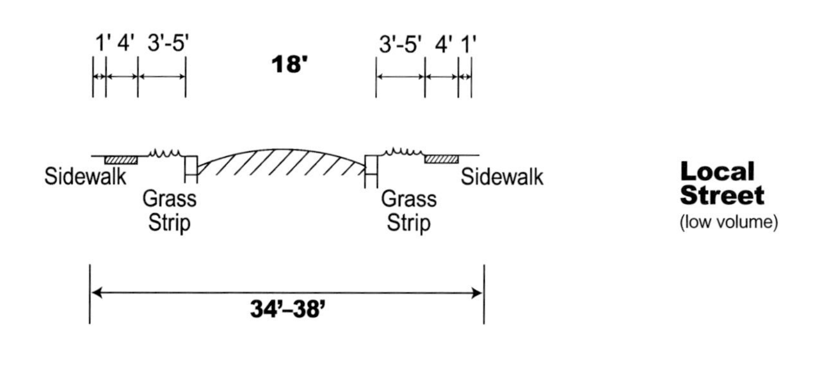

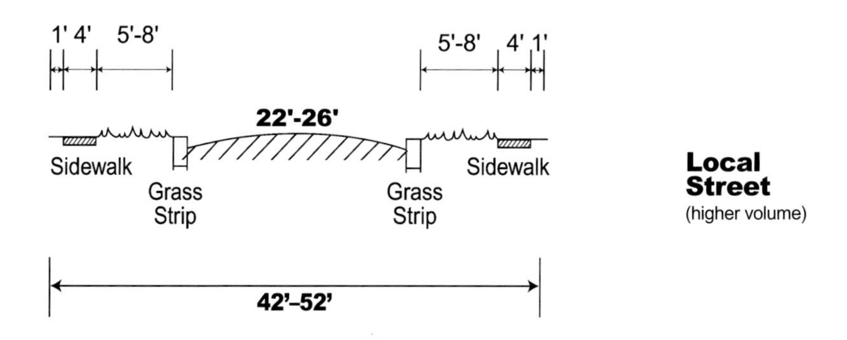

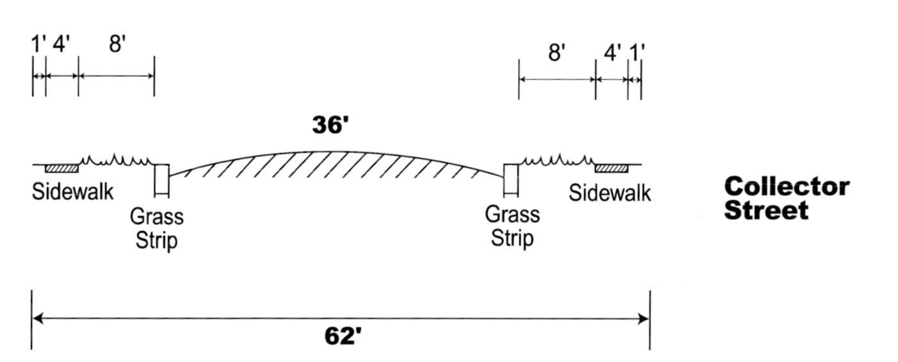

GLATTING JACKSON KERCHER ANGLIN LOPEZ RINEHART

parking bays provided at the edge of the street (Figure 2-17). Marked parking lanes, which usually are appropriate only on residential collector streets, require an eight-foot paved width. On unmarked local and access streets, parked vehicles occupy a space seven feet wide.

On streets without curbs, an eight-foot improved shoulder can be used in lieu of paved parking lanes. Such shoulders can reduce the rate of stormwater runoff and may also help create a natural or rural appearance. But material displaced

FIGURE 2-17
On-street parking bays.

from improved shoulders can be a nuisance for the paved area. Roadside shoulders need to be sensitively landscaped to meet functional and aesthetic objectives. In addition, they need to be carefully designed and constructed to ensure permanence and to avoid excessive maintenance requirements.

Where continuous on-street parking is impossible—such as on steep cross-slopes or where saving trees is an issue—angle parking at selected locations along the street may be suitable. Angle parking, however, requires not only the depth of the parking spaces but potentially also more moving-lane space than does parallel parking.

Off-Street Parking

The availability of off-street parking minimizes the need for parking lanes on the street. Off-street parking also keeps the street clear for snowplowing, emergency vehicles, delivery vehicles, and other wide vehicles. For these reasons, community designers should provide sufficient off-street parking as an alternative to curb parking. Residential occupant parking should be accommodated off street by driveways, carports, and garages, or by parking lots in higher-density developments (Figure 2-18). Because about one-third of all American families

FIGURE 2-18
Residential parking should be accommodated off the street by driveways, carports, and garages or by parking lots in higher-density developments.

BOB DUNPHY

own two or more cars, parking for residents is a major consideration in residential design. Automobile ownership per dwelling is a function of several variables, including family size and ages, family income level, wage earners per family, population density per acre, the availability of adequate public transportation, and proximity to schools and shopping centers. Determining what amount of off street parking is sufficient requires independent investigation of local conditions in light of current demand and reasonable allowance for future trends in automobile ownership.

Alleys

Alleys provide alternative vehicular access to homes and offer several benefits. They can be particularly effective in reducing driveway entrances to the street. Locating garages and driveways at the rear of properties improves the streetscape by eliminating the sight of cars parked in driveways and avoiding house designs that present the garage as the dominant feature seen from the street. Further, the reduction of driveway entrances to the street opens up street space that can be used more efficiently for additional parking needs.

Alleys are most often associated with older, densely settled urban areas, but they are used also in new suburban developments of single-family detached and attached houses. When subdivisions are designed with lot widths of less than 50 feet, alleys should be considered (Figure 2-19).

To use land efficiently, alleys should consume no more area than is necessary for the passage of a single service vehicle. A 12-foot pavement width with a 16-foot right-of-way will easily accommodate the widest of truck bodies (eight feet) with room to spare on both sides.

HNTB

FIGURE 2-19
Alleys tend to be associated with older, densely settled urban areas, but they are used also in new suburban developments of single-family detached and attached houses.

Alleys should not be constructed with curbs, the presence of which would necessitate a wider road surface to permit opposing vehicles to pass. Instead of curbs, planners should consider a two-inch invert in the cross-section of the alley pavement for stormwater runoff. Garages and fences should be set back from the alley right-of-way by three to five feet to provide an adequate turning area for vehicles. With the increased use of narrow lots and a renewed concern for the streetscape, alleys may enjoy a renaissance in the years ahead.

Design Speed

Design speed is the speed selected by planners for determining the various geometric design features of a planned roadway. Some of the design features of streets that may affect speed include the following:

- Open width or clearance of the street. A street with wide lanes invites higher speeds, as does a street with no visual obstructions within the right-of-way, enabling drivers to see over long distances.
- Horizontal curvature. The longer the radius of a curve, the higher the speed through that curve.
- Sight distance. The straighter and more level that streets are, the longer their sight distances. Long sight distances tend to encourage higher speeds.
- The number of access points to the street. The presence of many obvious potential conflict points on streets tends to inhibit speeding.
- The number of parked cars and other traffic-calming devices. The presence of parked cars and other potential obstructions effectively decreases traffic speeds.
- Signs and signals at controlled intersections. Traffic-control devices slow traffic within the immediate vicinity of controlled intersections.

Residential streets—because of their short lengths, the likelihood of parked vehicles, the presence of children and pets, and other reasons—should be de-

TABLE 2-5

DESIGN SPEEDS (in mph)

	Type of Terrain		
	Level	**Rolling**	**Mountainous**
Local Streets[1]	20	20	15-20
Residential Collector[2]	30	25-30	25

[1] Includes local streets in all ADT ranges. On local streets in mountainous terrain, 15 mph may be an acceptable design speed.

[2] Includes residential collector streets in all ADT ranges.

signed for low vehicle speeds. In addition, given the practical limitations and economics of street construction, if the terrain is difficult designers will lower design speeds. A design speed of 20 mph for all types of local streets is recommended (Table 2-5). By contrast, design speed plays a more important role in the design of residential collector streets, on which expedited movement is a more important function.

Gradients

Street grades in residential areas should be as flat as is consistent with the surrounding terrain. The gradient for local streets should be less than 12 percent except in unusual terrain, although a gradient of less than 8 percent is preferable. In mountainous terrain with fewer than 20 to 30 days of freezing temperatures, short segments of grades up to 18 percent can be accommodated. On residential collector streets, the gradient should be less than 10 percent and, under ideal circumstances, less than 7 percent. Where grades of 4 percent or steeper are necessary, the drainage design may become critical and special care must be taken to prevent the erosion of slopes and open drainage facilities. In areas with severe icing conditions, maximum grades of 8 percent may be preferable for all types of design conditions. (AASHTO Greenbook, 1990).

A minimum gradient on all curbed streets is necessary to prevent water from ponding. The minimum gradient is usually specified as 0.5 percent; however, it may be reduced to 0.35 percent in particularly flat terrain. If successful performance is to be achieved, subgrade compaction and construction gradient controls become critical when gradients fall below 1 percent. The design of all streets should also include a cross-slope to drain water off the traveled portion of the roadway. Pavement cross-slope is usually specified at about one-quarter inch per foot, but three-eighths inch per foot is recommended for streets that must accommodate short periods of intensive rainfall.

In mountainous terrain, there may be no practical way to avoid street gradients steeper than those considered maximum in less severe topography. Designers should devote particular effort to specifying the flattest possible gradients near intersections—for residential collectors, 3 percent maximum within 100 feet of an intersection; and for local streets, 5 percent maximum within 50 feet of an intersection. If unusually steep gradients are unavoidable, designers should provide for the following:

- both uphill and downhill access to every property, if at all possible;
- in snow country, additional parking at the foot of the hill; and
- pedestrian walkways with handrails or with flights of several steps located intermittently along the route.

Vertical and Horizontal Alignments

Recommended design values for safe-stopping sight distance are shown in Table 2-6. These sight distances should be provided on both horizontal and vertical curves.

TABLE 2-6

SAFE-STOPPING SIGHT DISTANCES

Design Speed (mph)	Stopping Sight Distance[1] (feet)
15	80
20	115
25	155
30	200
35	250
40	305
45	360

[1] Measured from driver eye height of 3.5 feet and object height of two feet.

Source: American Association of State Highway and Transportation Officials, *A Policy on Geometric Design of Highways and Streets* (Washington, D.C.: AASHTO, 2001), Exhibit 3-1. Used by permission.

Vertical Alignment

The vertical alignment of residential streets should allow for 1) grades that drivers can negotiate in adverse weather and 2) sight distances that are adequate for safety (Figure 2-20). Prescribed guidelines for maximum permissible grades are unreasonable, because individual situations dictate the appropriate interrelation between grading and drainage needs. Further, a community's aesthetic values, attitudes, customs, and design preferences play an important role in determining alignments. In climates with fewer than 20 to 30 days in which temperatures go below freezing, grades of up to 18 percent can be routinely accommodated on local streets. In colder climates, short segments of grades up to 12 percent and extended grades up to 8 percent can be accommodated.

Adequate sight distances on vertical curves is assured by providing a length of vertical curve (Figure 2-20) as indicated in Table 2-7.

FIGURE 2-20
Sight distance on the crest of a hill is the distance at which a driver can see an object two feet above the road.

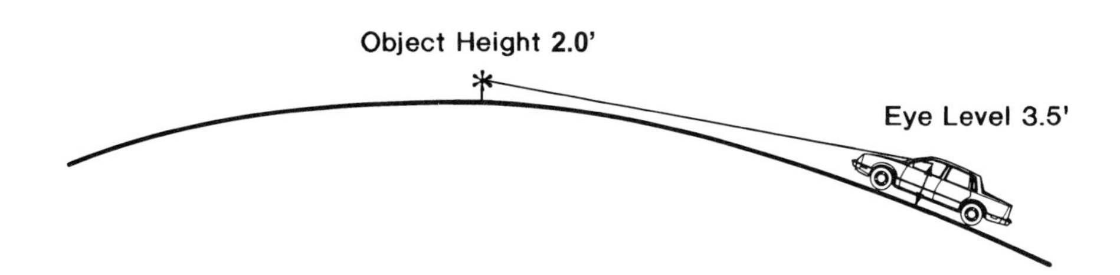

TABLE 2-7

MINIMUM RATE OF VERTICAL CURVATURE

Design Speed (mph)	Stopping Sight Distance (feet)[1]	Rate of Vertical Curvature K[2]	
		Crest Curves	Sag Curves
15	80	3	10
20	115	7	17
25	155	12	26
30	200	19	37
35	250	29	49
40	305	44	64
45	360	61	79

[1] Measured from driver eye height of 3.5 feet and object height of two feet.

[2] Rate of vertical curvature, K, is the length of curve per percent algebraic difference between intersecting grades.

Source: American Association of State Highway and Transportation Officials, *A Policy on Geometric Design of Highways and Streets* (Washington, D.C.: AASHTO, 2001), Exhibits 3-76 and 3-79. Used by permission.

Vertical curve length should be a minimum of 50 feet. When the algebraic difference in grades is less than 1 percent, no vertical curve is necessary for local residential streets.

Horizontal Curves

Horizontal curves on low-speed residential streets should not use superelevation (banking) to counteract centrifugal force on the vehicle; the design should allow centrifugal force to be counteracted solely with side friction from the pavement (Figures 2-21 and 2-22). Table 2-8 summarizes guidelines for the length of the centerline radius in horizontal curves. If centerline radii are increased, driver comfort will be enhanced. But frequently, the decision to lengthen curve radii has adverse consequences such as the loss of natural terrain, trees, and watercourses and an increase in vehicle speeds.

Dead-End Turnarounds

On most dead-end streets, a circular turnaround area is preferable (Figure 2-23). A T- or Y-shaped turnaround may be used for streets having a short length, alleys, and streets serving up to ten houses (Figure 2-24). Extremely short streets serving no more than five houses may not require a turnaround: vehicles may simply use the street width for backing/turning movements.

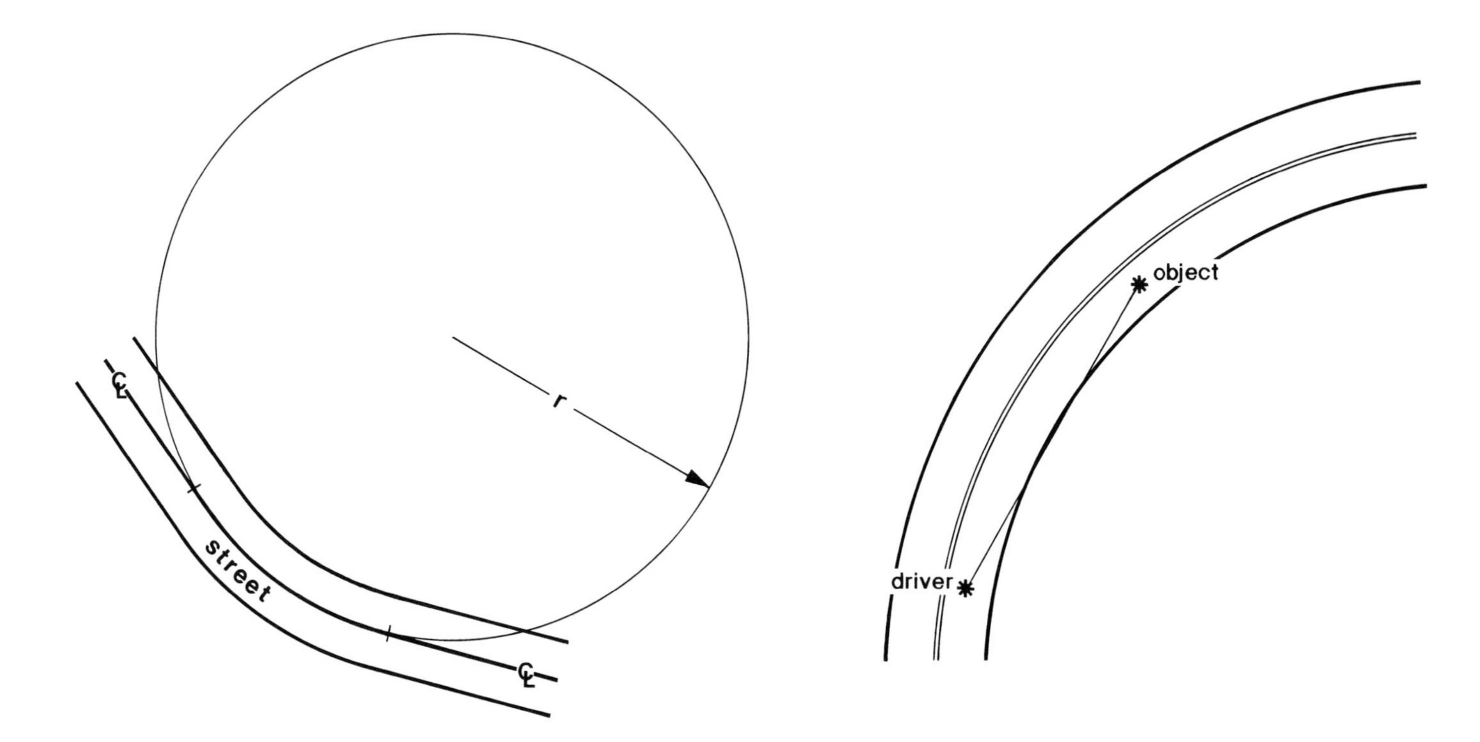

FIGURE 2-21
Horizontal curves can be described by the radius of the circle formed by the centerline of the curve (at left).

FIGURE 2-22
Sight distance on a horizontal curve is limited by the sharpness, or radius, of the curve (at right).

When a street that will be extended at a later date is constructed as a temporary dead-end street, a temporary turnaround must be provided. Curbs, if used, should be terminated before the entrance to the turnaround. Such dead ends generally should be barricaded to prevent vehicles from leaving the street.

The recommended radius for the paved area of a circular turnaround without a center island serving passenger vehicles is 30 feet (Figure 2-25). If frequent use of the turnaround by single unit vehicles (municipal services equipment, school buses) is likely, a 42-foot radius may be required. Single unit vehicles can use a turnaround with a 30-foot radius, but backing would be required. A 42-foot

TABLE 2-8

CENTERLINE RADIUS

Design Speed (mph)	Centerline Radius[1] (feet)
15	45
20	90
25	165
30	260
35	375
40	510
45	670

[1] Based on 0% superelevation for 15 to 25 mph, 1% for 30 mph, 2% for 35 mph, 3% for 40 mph, and 4% for 45 mph.

Source: American Association of State Highway and Transportation Officials, *A Policy on Geometric Design of Highways and Streets* (Washington, D.C.: AASHTO, 2001), Exhibit 3-40. Used by permission.

CLYDE DAVIS. COURTESY OF MOBILE COUNTY (ALABAMA) PUBLIC WORKS

BOB DUNPHY

FIGURE 2-23
Circular turnarounds.

radius can accommodate SUVs and other large passenger vehicles as well as all commercial and service vehicles with a regular need to visit residential streets, including school buses, all types of delivery trucks, emergency vehicles, solid waste collection trucks, and repair services vehicles. An off-center turnaround creates visual variety and makes turning easier (Figure 2-26).

To reduce the amount of paving, turnarounds may have center islands. Such turnarounds should provide adequate maneuver space—a minimum of 18 feet—around the island (Figure 2-27a). One option for expediting turning movements is to make the street pavement wider at the rear of the center island (Figure 2-27b).

The use of T- or Y-shaped turnarounds requires all vehicles to make a backing-up movement. This inconvenience can be justified if traffic volume on the turnaround is particularly low, as on a dead-end street serving ten or fewer homes with driveways. T- or Y-shaped turnarounds offer several advantages:

FIGURE 2-24
A T- or Y-shaped turnaround may be used for shorter streets with up to ten houses.

NAHB

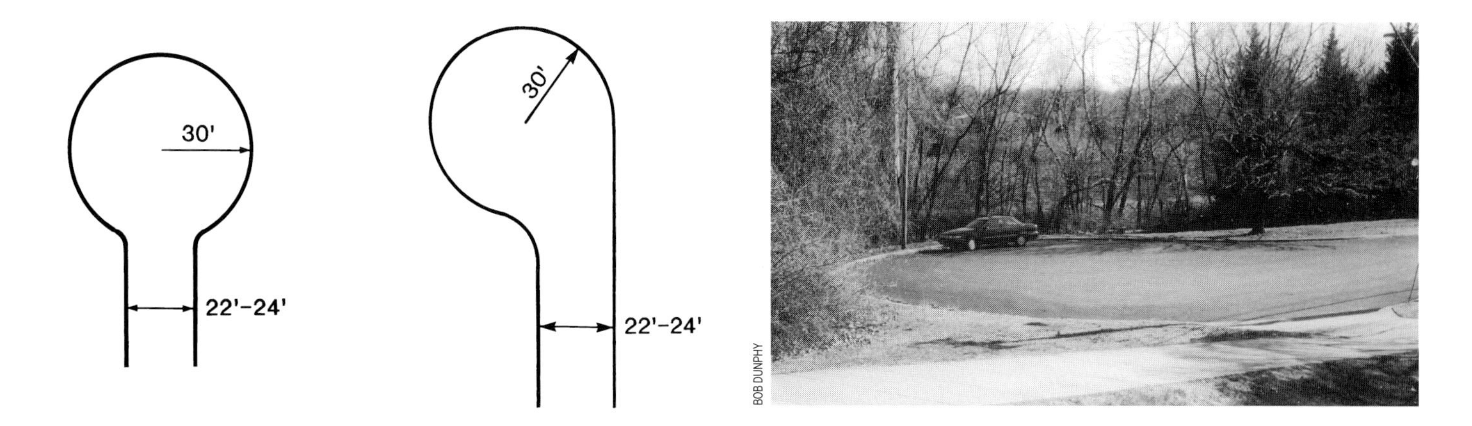

FIGURE 2-25
A 30-foot radius is recommended for a circular turnaround (drawing at left).

FIGURE 2-26
Off-center turnaround (drawing at right and photograph).

they require less paving, their construction and maintenance costs are lower, and they provide greater flexibility in land planning and the location of homes. The recommended dimensions for T- or Y-type turnarounds are 60 feet by 20 feet (Figure 2-28), yielding a paved area only 43 percent as large as the smallest (30-foot radius) circular turnaround. While T- and Y-shaped turnarounds are limited in their applications, they are an important option for particularly low volume streets.

Another turnaround option is the auto court, which is a courtyard surrounded with parking. Attractive landscaping and paving can make an auto court a neigh-

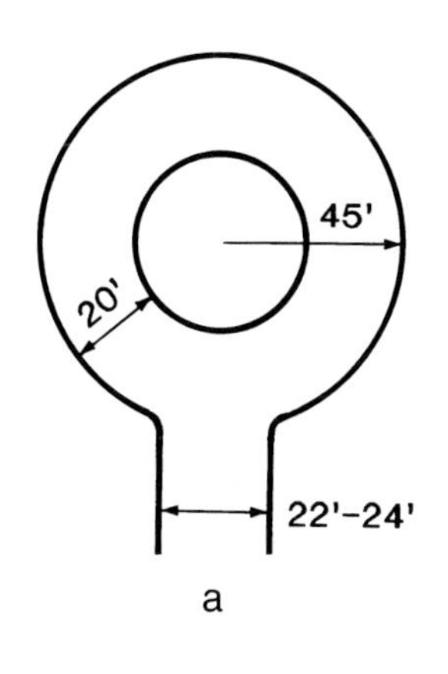

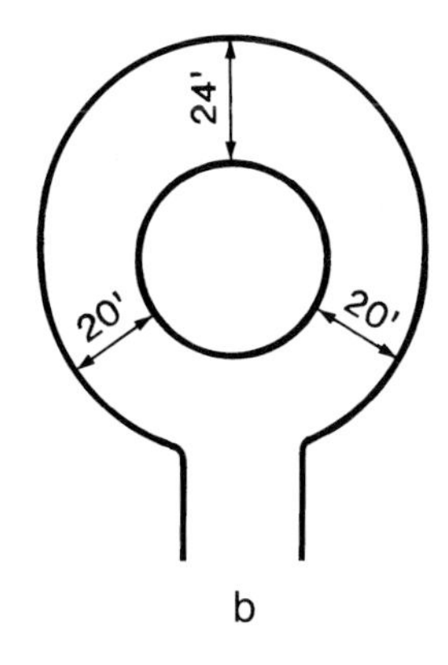

FIGURE 2-27
Circular turnarounds with center islands.

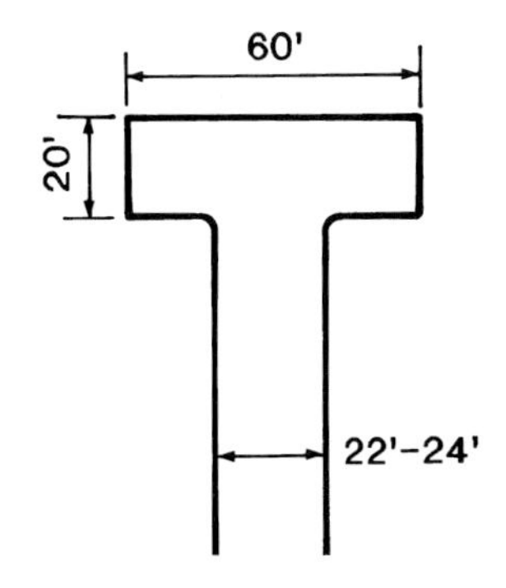

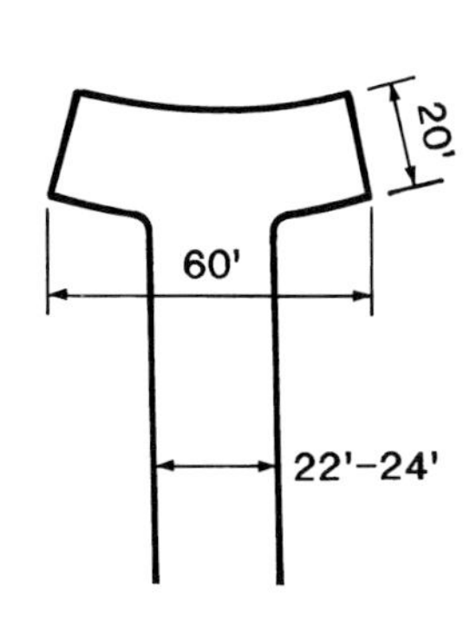

FIGURE 2-28
T- and Y-shaped turnarounds.

CLYDE DAVIS, COURTESY OF MOBILE COUNTY (ALABAMA) PUBLIC WORKS

NAHB

FIGURE 2-29
Auto court (at left) and eyebrow (at right).

borhood amenity (Figure 2-29). Another possibility is a close (or eyebrow street), which is a small loop that functions as an access street and parking area and usually incorporates landscaped islands.

Parked vehicles on any turnaround will reduce available turning space. Parking restrictions, signing, and enforcement may at times be desirable.

The recommended right-of-way for circular turnarounds is a radius ten feet greater than the paved area. Therefore, a paved radius of 30 feet should have a 40-foot right-of-way radius and a radius of 42 feet should have a 52-foot right-of-way radius. For T- or Y-shaped turnarounds, the right-of-way should extend 10 feet from the edge of the street pavement.

Dead-End Street Length

The longer dead-end streets get, the more isolated and difficult to reach because the properties along them are accessible from only one direction. Lengthy dead-end streets begin to assume the function of higher-order streets (local or residential collector streets) that provide access to more than a few properties.

In general, traffic volume and, correspondingly, the number of housing units should be factors that determine the length of dead-end streets. In general, a dead-end street should handle no more than 200 vehicle trips per day. Assuming that each single-family house generates eight to ten vehicle trips per day, a dead-end street should accommodate a maximum of 20 to 25 houses.

Shared Driveways

One alternative method of serving a few homes is shared driveways. A shared driveway is a paved access that is not built to street standards, but is simply a normal residential driveway that extends and branches off to several houses. Privately owned and maintained, shared driveways do not require a turnaround area at the end; instead, they simply end at the last house. Shared driveways can connect to an access street or local street at a right angle or, as an extension of a dead-end street, they can connect to the dead-end street turnaround

HNTB

FIGURE 2-30
An economical and attractive method of serving up to five or six houses, shared driveways should be just wide enough for two cars to pass.

(Figure 2-30). An economical and attractive method of serving up to five or six houses, shared driveways should be just wide enough for two cars to pass, which is approximately 16 feet.

Streetscape

Residential streets should provide not only safe, efficient circulation for vehicles and pedestrians, but also should create positive aesthetic qualities for the residents and the community. Paying attention to the aesthetics of streets helps assure that they do not become simply thoroughfares for vehicles. The character of a residential street is influenced to a great extent by its paving width, its horizontal and vertical alignments, and the landscape treatment of the street edges. Residential streets are community spaces that should convey an image and scale appropriate to the neighborhood. For example, much of the character of older neighborhoods is derived from the mature street trees that form a canopy over entire streets. By contrast, a neighborhood with wide streets devoid of trees conveys an entirely different image. The placement of utilities and the style of traffic-control devices and street-lighting hardware also contribute to the character of the street (Figure 2-31).

Trees and Shrubs

When planning a streetscape that is to include trees and shrubs in the right-of-way, a number of important factors must be considered.

Choice of Plants. The choice of tree species is a critical decision that should be left to an urban forester, arborist, or landscape architect. When selecting street trees, the consultant should consider their mature height and spread, the root system's potential for damaging sidewalks and street pavements, maintenance requirements, tolerance to pruning, and adaptability to the specific street

DAVID JENSEN ASSOCIATES, INC.

HNTB

FIGURE 2-31
Streetscape treatments. Landscaping and signs can help create an attractive streetscape. Trees and shrubs are planted in the right-of-way in Celebration, Florida (at bottom).

DAVID JENSEN ASSOCIATES, INC.

THE WALT DISNEY COMPANY

environment. The trees selected should have branches that are high enough to allow commercial vehicles to pass underneath them.

Shrubs selected for right-of-way planting should be low growing or, in the case of large shrubs, tolerant of undertrimming. Low shrubs and ground covers with vigorous root systems can be effective for erosion control on slopes within the right-of-way.

Before planting trees or shrubs near streetlights, the consultant should check their photosensitivity. Continuous exposure to streetlights can cause abnormal growth in certain trees and shrubs. In northern states where salt is used on the roadways for snow and ice removal, salt-tolerant species should be selected.

Location of Plants. All shrubs and trees planted at intersections and driveways should permit a clear sight distance in a two- to eight-foot area above the street, providing visibility for cars and trucks. Periodic pruning can maintain the required visibility. Street trees should be planted at least three feet behind the back of the curb.

Preservation of Existing Vegetation. On new projects, it is often possible to incorporate existing trees and shrubs into the streetscape by carefully planning alignments and grades. Existing trees can be saved by curving roads around them or by creating islands that divide streets into one-way pairs. Vegetation need not be cleared from the entire right-of-way, but only as needed to accommodate utilities, sidewalks, and drainage, thereby preserving as many trees as possible.

Maintenance Requirements. Trees and shrubs should be selected for their low maintenance requirements. Consideration should be given also to the amount of leaf and fruit litter. The responsibility for the maintenance of trees and shrubs in private-drive easements should be clearly defined.

Utility Placement

Whenever possible, utilities should be placed underground. Certain elements that must be located aboveground—for example, transformers and junction boxes—can be camouflaged by painting them with colors that blend into the background or by screening them with plant materials. The location of aboveground utility structures should be coordinated with the proposed landscape plans, to ensure that they are positioned away from key focal areas. The light poles and fixtures selected could relate to the architectural styles and character of the housing in the neighborhoods they illuminate. Most utility companies offer a selection of streetlamp and pole styles that meet the needs of most neighborhoods.

Signs

The *Manual on Uniform Traffic Control Devices for Streets and Highways* (M.U.T.C.D.), issued and regularly updated by the U.S. Department of Transportation, Federal Highway Administration, is the definitive source for standards on street signs within residential neighborhoods. In contrast to geometric design material from, for example, ITE and AASHTO, the sign directives contained in the *Manual on Uniform Traffic Control Devices* are standards. Unlike guidelines, these standards are not permissive of variation or judgment. To the contrary, there are strong legal and liability-related reasons for adhering strictly to the requirements of the *Manual on Uniform Traffic Control Devices*.

For example, M.U.T.C.D. recognizes these types of signs appropriate for residential streets:

- Regulatory Signs. These signs (for example, "Stop," "Yield," "Speed Limit 15 mph," "Do Not Enter") control the operation of motor vehicles. Motorists' disregard of these signs is a violation of motor vehicle laws.
- Warning Signs. These signs (for example, "Curve," "Turn," "Winding Road") warn motorists of an upcoming situation in which driving care is advised. Typically, these signs are diamond shaped with black text on a yellow background. Motorists' nonobservance of warning signs is not a driving law infraction.
- Guide Signs. These signs (for example, destination and distance indicators, street names, highway numbers) provide location and route information to drivers. Some design individuality is allowed in these signs. However, the recommendations of M.U.T.C.D. should be consulted in order to maintain the consistency of guide signs within areas.
- Recreational and Cultural Interest Signs. Such signs are used to identify the locations of recreational areas, cultural sites, historic landmarks, and other places of interest. If such destinations are located within a community, it is appropriate to identify them through these signs.

M.U.T.C.D. should be reviewed for any traffic-control devices that are under consideration for use in residential communities, and the manual's standards should be applied consistently.

Pedestrian and Bicycle Access

Today's emphasis on comprehensive community planning has made sidewalks, pedestrian pathways, and bicycle paths an integral part of residential land development. Successful residential street design now accommodates nonmotorized travel (pedestrians and bicycles) at a level of attention comparable to that given to vehicular flow.

Pedestrian Walks and Paths

Sidewalks not only provide a circulation network for pedestrians, but also can serve as a neighborhood meeting place and a play area for children. In general, sidewalks should be provided on both sides of the street. Typically they are five feet wide, and at a minimum they are four feet. In certain unusual conditions—for example, where side slopes are steep or a historic area exists—sidewalks on only one side of the street may be appropriate, and this may also be the case for

HNTB

FIGURE 2-32
A three- to five-foot border area or grass strip between the street edge of the sidewalk and the curb face is desirable in most residential areas.

large-lot developments where frontages exceed 150 feet. Wider sidewalks may be considered in locations next to significant pedestrian generators such as schools, businesses, high-density dwellings, transit stops, and churches.

Paths and sidewalks should always be located within a public right-of-way, a public easement, or a common area. A common local guideline suggests that sidewalks should be located at least one foot inside the right-of-way line or easement if no aboveground utilities present conflicts.

A three- to five-foot border area or grass strip between the street edge of the sidewalk and curb face is desirable in most residential areas (Figure 2-32). This grass border provides a visual break between the paved surfaces of the street and sidewalk, as well as the following benefits.

- Children walking and playing enjoy increased safety from street traffic.
- The border can be used for temporary storage of trash receptacles awaiting pickup at the edge of the street, eliminating conflicts between pedestrians and trash/recycling containers.
- The sloped transition area necessary for an appropriate driveway gradient across sidewalks can be minimized by locating a major portion of the gradient within the border.
- The danger of collision between pedestrians and out-of-control vehicles is minimized by the placement of sidewalks at the maximum practical distance from the curb.
- Snow plowed off roadways can be stored on the border, leaving sidewalks open for pedestrians.
- In rainy weather, pedestrians are less likely to be splashed by passing vehicles.
- Borders provide planting space along streets.

Providing a grass strip between streets and sidewalks may be difficult and inadvisable in demanding terrain, particularly along streets on a steep cross-slope.

USED WITH PERMISSION OF KEVIN BESWICK/PEOPLE, PLACES, AND THINGS PHOTOGRAPHIC. DAVID JENSEN ASSOCIATES, INC.

FIGURE 2-33
A winding path through the neighborhood provides an attractive route for persons in wheelchairs, walkers, bikers, and skaters.

In arid regions where an irrigation system would be needed to sustain grass, a hardscaped strip rather than a planted strip may be appropriate. Along collector streets bordered by schools, churches, businesses, and other travel destinations, foot traffic may be considerable and landscape maintenance a problem. In such areas, the installation of a continuous six- to eight-foot sidewalk without a grass strip may be appropriate.

Paved paths that may not strictly follow the alignment or grade of streets offer an option to traditional sidewalks. In developments with steep topography and narrow lot widths, frequent changes of grade from sidewalks to driveways may cause difficulties for people in wheelchairs. Under these conditions, providing alternative walkways may be preferable. These may be slightly winding paths located within the street right-of-way or paths far removed from the street that weave their way through neighborhoods (Figure 2-33). Paths not within street rights-of-way are usually located in common areas that are usually maintained by a homeowners association.

Bicycle Routes and Paths

Bicycle travel has taken its place along with pedestrian travel as an important consideration in residential street planning, warranting the same level of design attention as motor vehicle travel. The rapidly increasing amount of attention being paid to the development of regional, public bicycle routes has made connections to regional systems valuable to individual subdivisions; and private community connections to regional routes have extended the usefulness of these routes. Purposeful bicycle riding, that is, riding not just for recreation but to accomplish a travel need, was almost nonexistent in the 1960s and 1970s and has been growing slowly but steadily for the past few decades. A number of residential communities accommodate purposeful bicycle travel in order to reap a sales advantage.

The backbone of a good bicycle network in a community is a well-connected network of local streets. Good residential street design in itself—a complete network of streets affording a variety of routes within the subdivision, with vehicle speeds contained through the design of the streets, and reasonably direct connections to adjacent areas—accommodates bicycle travel as well as motor vehicle travel.

If properly laid out, a subdivision's entire network of local and access streets will provide an attractive setting for bicycle travel. No special modifications need be made to well-designed local and access streets in order to fully accommodate bicycles on those streets.

On residential collector streets within subdivisions, however, it may be appropriate to add on-street bicycle lanes. Conditions that support the inclusion of an on-street bicycle lane include the following.

THE WALT DISNEY COMPANY

FIGURE 2-34
If properly laid out, a subdivision's entire network of local and access streets will provide an attractive setting for bicycle travel (at left). On-street bicycle lane guidelines (below).

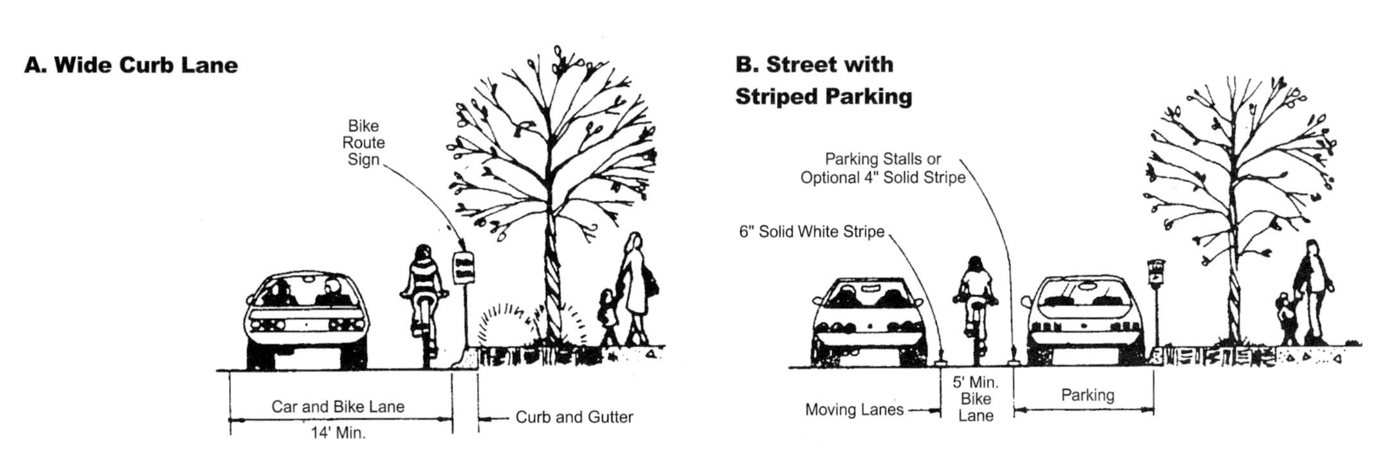

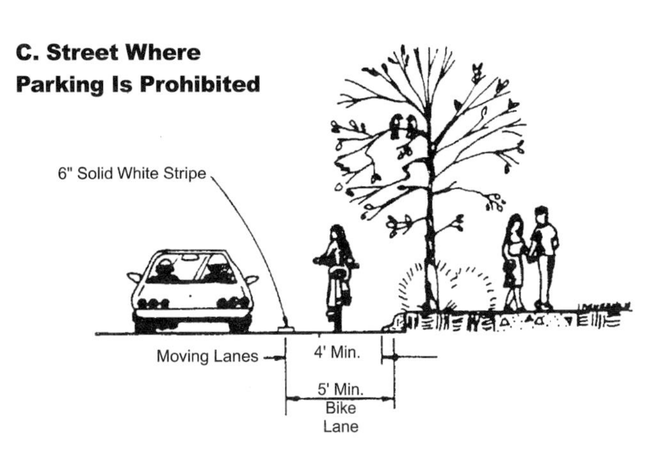

GLATTING JACKSON KERCHER ANGLIN LOPEZ RINEHART

- The residential collector street carries a significant portion of the development's total traffic.
- The network of local streets is incomplete, forcing bicycle travel onto the collector street as the only available route through the community.
- Destinations for purposeful bicycle travel—such as a school, a college campus, recreational facilities, or a business park—are located nearby.
- The subdivision is able to connect to an exterior system of bicycle trails and lanes and the subdivision's bicycle lanes represent important links for the regional system.

AASHTO has issued a report, *Guidelines for the Development of Bicycle Facilities* (1991), that provides guidance for the design of on-street bicycle lanes (Figure 2-34). These guidelines recommend a five-foot striped bicycle lane at the outer edge of the roadway or, where on-street parking is present, between the parked vehicles and the moving traffic lane.

The need for off-street bicycle paths is a function of the subdivision's density, motor vehicle traffic volumes on nearby streets, residents' preferences, and the proximity of bicycle-trip generators, such as educational institutions or parks. Planners should also consider the location of the newly developing area in relation to any established community-wide bicycle path system.

Path sharing by bicycles and pedestrians is appropriate, particularly if the paths loop through the subdivision and are not used by through traffic, meaning bicyclists or pedestrians making trips that begin or end up outside the subdivision. For a shared path, the paved area should be ten feet wide or eight feet at a minimum. On wide trails, pavement striping can assist in avoiding conflicts between bicyclists and pedestrians

Bicycle path and sidewalk street crossings should be located at points that offer adequate sight distance. Curb cuts should be provided for bicycles, wheelchairs, baby carriages, and other wheeled vehicles. When heavily traveled paths cross busy residential collector streets, safety devices such as signs, signals, and painted crosswalks should be used. When paths intersect arterials and heavily used collector streets, bicycles and pedestrians should cross by means of special signal controls or overpasses.

Curb Cut Ramps

The construction of new sidewalks in residential subdivisions or their repair and replacement should follow the design guidelines issued in the 1990 Americans with Disabilities Act (ADA). The guidelines are summarized in numerous publications, including the recent *Accessible Rights-of-Way: A Design Guide* (U.S. Architectural and Transportation Barriers Compliance Board and U.S. Federal Highway Administration, November 1999). The following principles (Figure 2-35) are from the ADA:

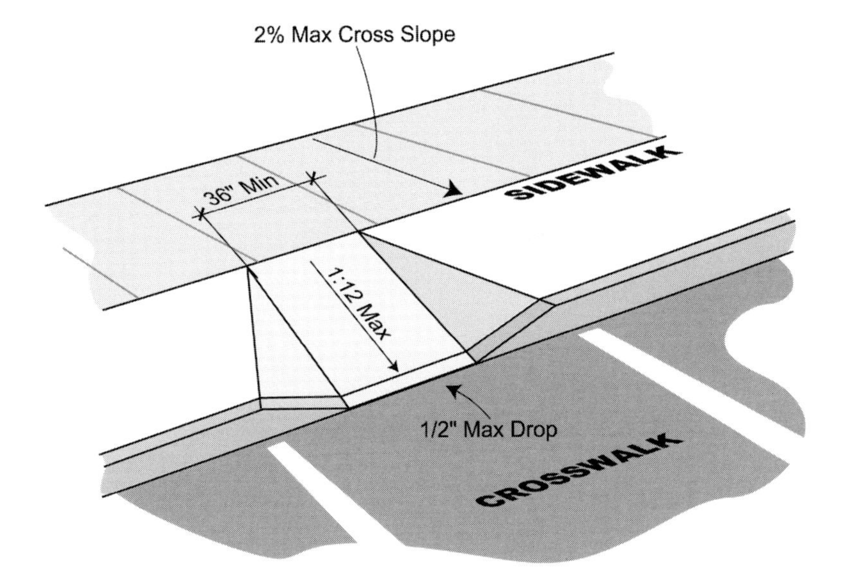

FIGURE 2-35
Americans with Disabilities Act requirements for curb cuts.

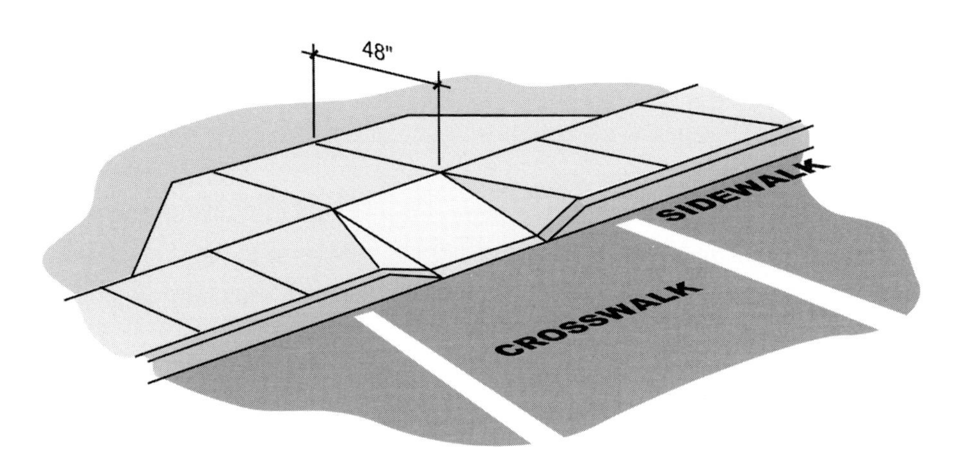

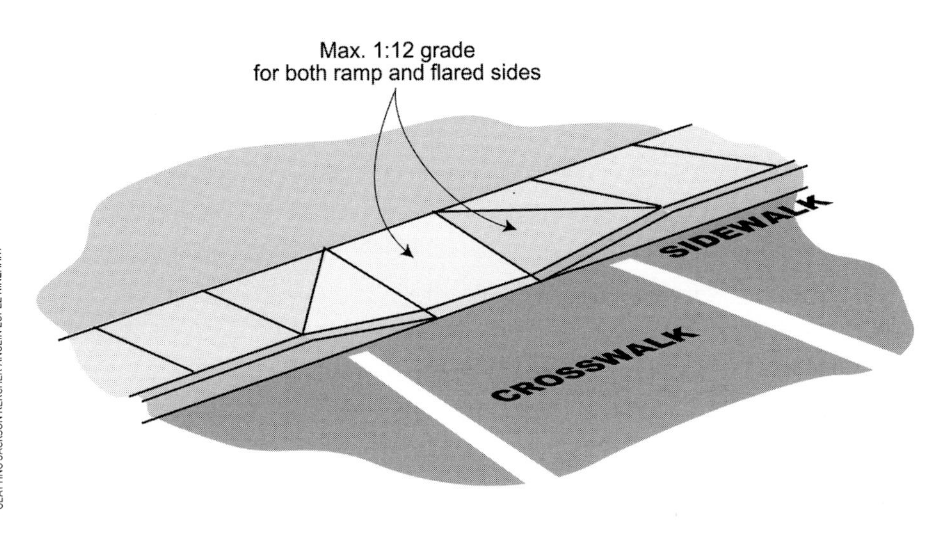

GLATTING JACKSON KERCHER ANGLIN LOPEZ RINEHART

FIGURE 2-36
Curb cut ramps should be located close to the intersection to keep the width of the crosswalk to a minimum.

- Curb cut ramps should be located close to the intersection to keep the width of the crosswalk to a minimum (Figure 2-36). Sidewalk furniture such as mail boxes and newspaper dispensers should not be allowed to block ramp areas.
- Paired curb ramps are preferred by users because such a configuration requires all pedestrians to enter a crosswalk at the same point and they provide more useful information to blind pedestrians about the location of the corner and the crossings.
- One ramp at the apex of the corner will generally suffice if it is wide enough to accommodate both directions of the streets.
- The surface of the ramp should be more textured than the surrounding sidewalk. Obtained by coarse brooming or scoring transverse to the slope of the ramp, a rougher texture provides a nonslip surface and helps warn sight-impaired pedestrians of the ramp.
- Care should be taken to ensure a uniform grade on the ramp, free of sags and short grade changes. The cross-slopes of ramp surfaces should not exceed one-quarter inch per foot

Curbs

Curbing is used to control drainage, protect pavement edges, and protect sidewalks and lawns from encroachment by vehicles. Curbs come in two general categories: vertical and sloping. Each kind of curb has many specific design variations and the major difference between vertical and sloping curbs is whether they restrain vehicle access (vertical curbs) or permit vehicle access (sloping curbs).

Vertical Curbs. Ranging from six to eight inches high and having steep sides, vertical curbs are relatively high and are designed to discourage vehicles from leaving the roadway (Figure 2-37). Vertical curbs should be placed at least one foot and preferably two feet from the edge of the travel way. Among the advantages of vertical curbs are the following:

DPZ MID-ATLANTIC

TDA INC.

BOB DUNPHY

FIGURE 2-37
Vertical curbs.

- They better protect pedestrians, street trees, utilities, and signs.
- They establish a positive limit of vehicle encroachment on the border area, minimizing parkway erosion and reducing the probability of vehicles sliding off the roadway under unfavorable pavement and weather conditions.
- They provide excellent drainage control.
- They are better able to control parked vehicles that on their own start to roll (runaways).
- They protect the border grass from damage by snowplows.

Sloping Curbs. Sloping curbs are designed so that vehicles can cross over them if necessary (Figure 2-38). Among the advantages of sloping curbs are the following:

- They allow subdividers and developers to undertake driveway construction without curb depression. Since driveway locations do not have to be determined before curb installation, developers enjoy some flexibility in the timing and location of driveway construction.
- They can accommodate off-pavement parallel parking.

CLYDE DAVIS, COURTESY OF MOBILE COUNTY (ALABAMA) PUBLIC WORKS

CLYDE DAVIS, COURTESY OF MOBILE COUNTY (ALABAMA) PUBLIC WORKS

FIGURE 2-38
Sloping curbs.

Sloping curbs that adjoin driveways must be designed so that the angle and height of the curb permit the passage of cars over the curb without causing the bottom of the car to scrape on the curb (Figure 2-39).

Asphalt curbs can be a useful alternative in low-traffic areas (Figure 2-40). Their chief advantage is their lower initial cost. However, they are less durable and require more maintenance than concrete curbs.

In relatively low-density developments, curbs may not be necessary at all. Factors that must be considered in deciding whether to build curbs are the method of handling stormwater, the expected amount of on-street parking, the cost of curbs, and the desired appearance of the community (Figure 2-41). Curbs and gutters are an important part of a community's stormwater drainage system, and their roles in handling runoff are discussed in greater detail in Chapter 4.

Curbless streets work best in rural estates where density does not exceed one home for every one or two acres. As residential density increases, the disadvantages of curbless streets are likely to start outweighing the advantages. Density increases mean more dwelling units and shorter driveways, increasing the

FIGURE 2-39
Improperly designed sloping curbs.

NAHB

NAHB

CLYDE DAVIS. COURTESY OF MOBILE COUNTY (ALABAMA) PUBLIC WORKS

DAVID JENSEN AND ASSOCIATES, INC.

need for on-street parking. Even a small amount of on-street parking along curbless streets is likely to pose problems—including physical damage to the roadside unless it has been prepared for parking, damage to the unprotected pavement edge of the road (raveling), and people parking in front yards and other areas off the road. All of these problems are effectively addressed by a street design with curbs.

FIGURE 2-40
Asphalt curbs can be a useful alternative in low-traffic areas (at left).

FIGURE 2-41
Curbless street (at right).

Traffic Calming

Since the mid-1990s, traffic-calming actions have become an important part of residential street design. Traffic calming is a combination of measures—mainly physical measures—undertaken to reduce the negative effects of motor vehicle use, alter driver behavior, and improve conditions for nonmotorized street users.

Most traffic calming is achieved in one of three ways: 1) narrowing the width of streets or their apparent width to drivers, 2) reducing sight distances with curves, and 3) adding texture to the driving surface. Typically, traffic-calming actions are applied to existing streets, usually to remedy problems stemming from the original design of the streets. Properly designed new residential streets contain a large number of traffic-calming features as part of their design and, therefore, should require little additional traffic calming once they are in operation.

The most important traffic-calming feature—narrow (or seemingly narrow) street widths—can be assured by simply selecting an appropriate mix of residential streets. Local streets—the backbone of the residential street system—feature minimum reasonable pavement widths and generally require that the pavement be shared by vehicles in both directions whenever parked vehicles are present. Pavement sharing is perhaps the single most powerful traffic-calming measure that can be designed. Other elements of good street design can further reinforce the sense of a narrow street, including curb and gutter drainage, uniform plantings of street trees as close as possible to the pavement edge, and the occasional presence of parked vehicles.

TABLE 2-9

CITIES WITH NARROW ROADWAY WIDTHS

The cities listed below are among communities in the United States that have successfully adopted narrow street standards.

	Total Width (feet)	Parking	Effective Width (feet)
Albany, Oregon	28	One side	20
Beaverton, Oregon	28	Both sides	12
Birmingham, Michigan	26	Both sides	10
Burlington, Vermont	18	None	18
Denver, Colorado	20[1]	One side[1]	12
Eugene, Oregon	21	One side	13
Forest Grove, Oregon	26	Both sides	10
Helena, Montana	33	Both sides	17
Loomis, California	24	One side	16
Mountain View, California	27	Both sides	11
Phoenix, Arizona	28	Both sides	12
Phoenix, Arizona[2]	36	Both sides	20

[1] Paired 20-foot streets with eight-foot median. Parking on one side of each roadway.

[2] Collector streets.

Sources: Livable Oregon, Narrow Streets Database; and TDA Inc.

Deflection of the vehicle path to decrease speeds is a method of traffic calming that can be effectively obtained through the original design of the streets. Streets designed with curves and also with grades following the natural contours of the terrain will have reduced sight distances, which works to lower the design speeds of such streets to acceptable levels. Intersection circles and roundabouts also reduce the sight distance for streets at intersections. Installing textured pavement at important areas, such as on pedestrian crosswalks, in business districts, and at school crossings, is a traffic-calming measure that can be built into the original design of the community.

A number of other traffic-calming measures that, typically, are associated with retrofits to older communities can be used in the design of new residential streets as well. The periodic narrowing of short sections of streets for a few car lengths near intersections or even at mid-block locations is an effective measure for traffic calming and is appropriate for new streets. The intermittent narrowing of streets to a single lane, forcing opposing vehicles to share the street even when no parked vehicle is present, is also an appropriate measure for new residential streets. Occasional chicanes (which are sharp bends created by curbing or barriers) constitute another traffic-calming device that would be appropriate for new subdivisions.

3

Intersections

Intersections are points of conflict and potential hazard. Therefore, the alignments of intersecting streets should be as straight as possible and their grades as flat as practical in order to afford drivers a complete and unobstructed sight distance, which enables them to make the necessary maneuvers to pass through intersections safely with a minimum of conflict between vehicles (Figure 3-1).

BOB DUNPHY

FIGURE 3-1
Intersection.

FIGURE 3-2
A three-legged T-intersection.

CLYDE DAVIS. COURTESY OF MOBILE COUNTY (ALABAMA) PUBLIC WORKS

Geometry of Intersections

The three-legged T-intersection is the simplest type of intersection (Figure 3-2). The automatic right-of-way assignment that is inherent in T-intersections, coupled with a significant reduction in conflict points, provides a high level of safety. The four-legged cross intersection is the most common type of intersection. Its repeated use produces a highly connected street pattern, the most efficient traffic distribution system there is (Figure 3-3). At four-legged intersections, stop or yield signs are usually necessary to establish the right-of-way. Although four-way intersections have potentially more conflict points than T-intersections,

FIGURE 3-3
Four-legged intersection.

CLYDE DAVIS. COURTESY OF MOBILE COUNTY (ALABAMA) PUBLIC WORKS

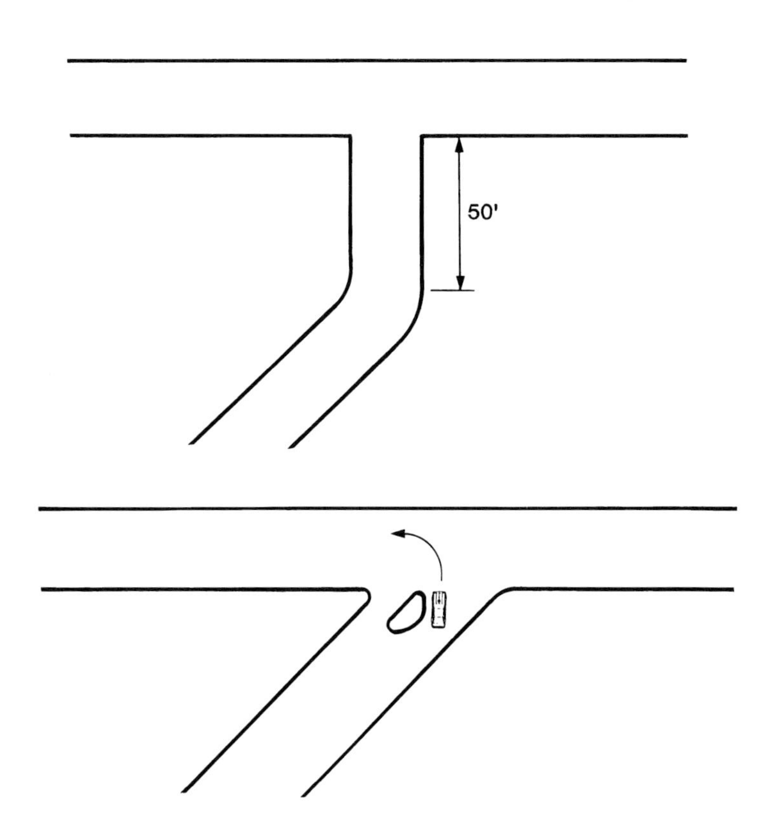

FIGURE 3-4
Realignment of an angled street to achieve a 90-degree intersection.

FIGURE 3-5
Landing area for an acute-angle intersection.

their convenience and traffic capacity have served many older communities well over the years.

Intersection Angle

In most cases, streets should intersect one another at as nearly 90-degree angles as possible. Right-angle intersections are the most comfortable for drivers and provide the most direct view of approaching traffic. It is awkward for drivers to gain an adequate view of traffic from acute-angle intersections. While local site conditions and land planning factors may require some intersections to be constructed somewhat off a right angle, the angle should be no less than 60 degrees.

When a street approaches another street at an undesirable angle, a better intersection can be created by bending a 50-foot section of the angled street to meet the other street at a 90-degree angle (Figure 3-4). Another method of compensating for a badly angled intersection is to install an island that separates traffic and channels the left-turn movement from the angled street into a nearly perpendicular turn (Figure 3-5).

Traffic Circles and Roundabouts

Traffic circles and roundabouts (Figure 3-6), both of which require traffic to proceed in a circular manner through an intersection, have been enjoying a resur-

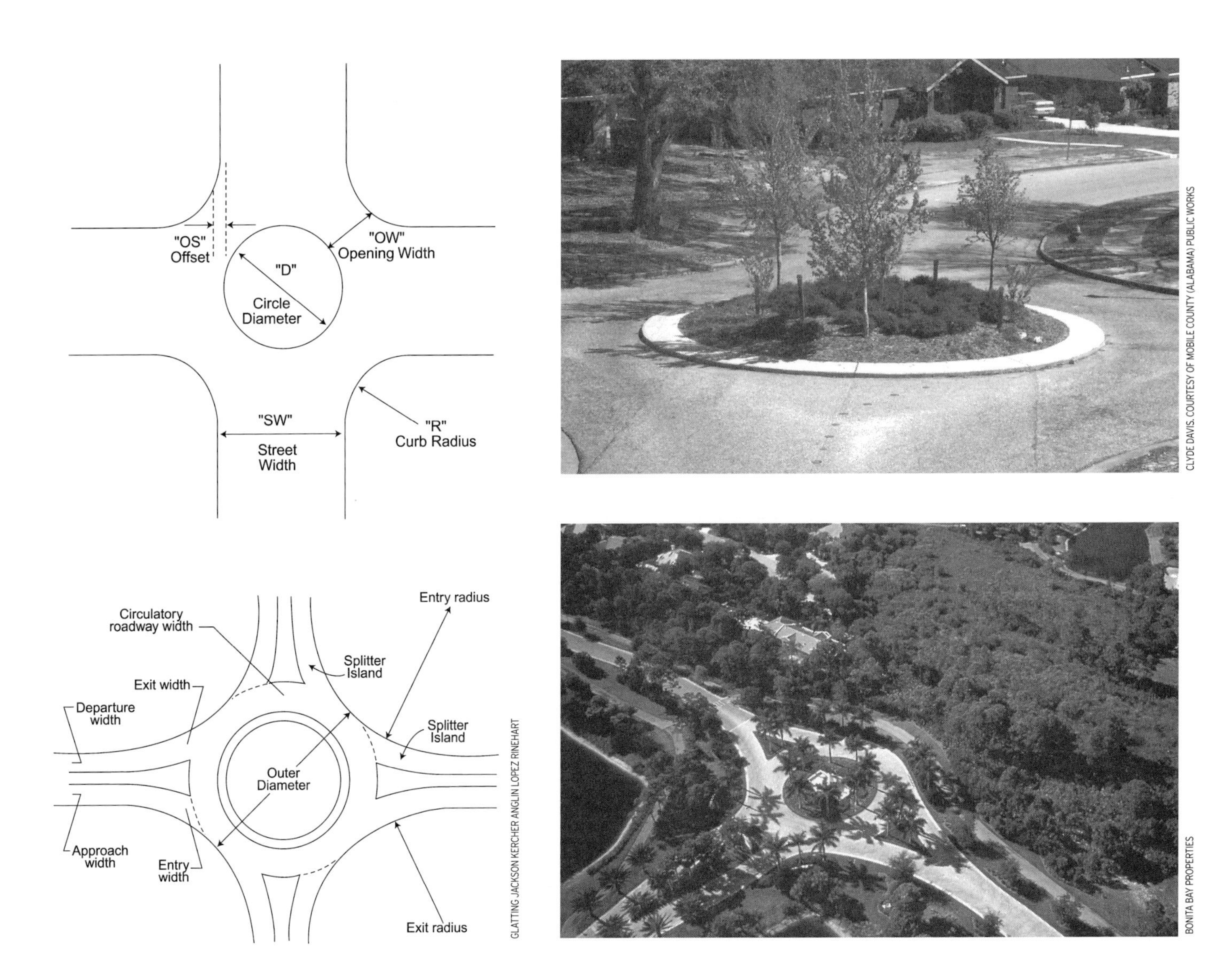

FIGURE 3-6
Traffic circles and roundabouts.

gence in popularity in residential street design. Intersection traffic circles on local streets are usually small—15 to 20 feet in diameter—and installing them in a normal four-way intersection requires no additional street space.

Roundabouts are generally larger than traffic circles, ranging from 25 to 40 feet in diameter. Typically, they require more street space than a normal four-way intersection. In a roundabout, the intersecting streets are flared as they approach the roundabout and triangular channelizing islands (called splitter islands) are centered in each connecting street. (Traffic circles do not include these features.)

Traffic circles are appropriate for the relatively low-speed, low-volume conditions on local streets. Roundabouts are more likely to be appropriate on higher-volume and higher-speed residential collector streets.

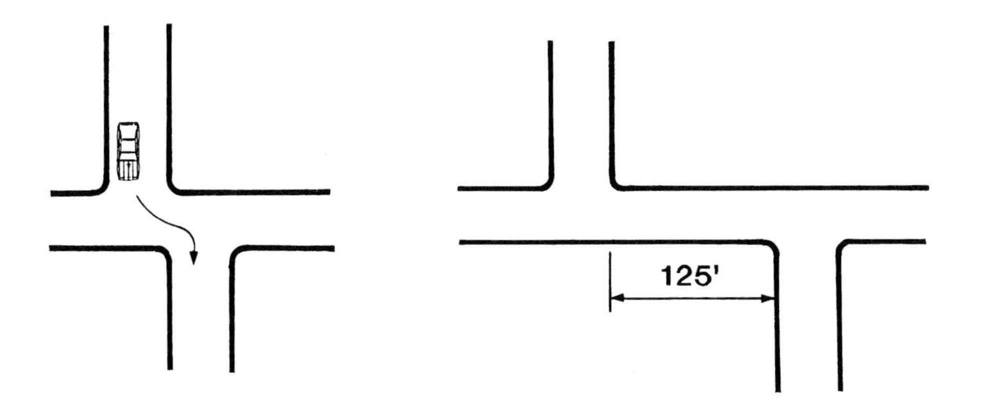

FIGURE 3-7
Corner cutting (at left).

FIGURE 3-8
A minimum separation of 125 feet between intersections is recommended (at right).

In traffic circles and roundabouts, traffic moves counterclockwise around the center island. Traffic within the intersection (that is, the traffic that is moving around the circle) has the right-of-way, and traffic entering from any direction yields right-of-way. At traffic circles, trucks, school buses, and other large vehicles make left turns by turning in front of the island rather than proceeding around it. Vehicles making this type of left turn yield right-of-way to approaching vehicles. At roundabouts, all vehicles proceed around the circle.

As a type of intersection control, traffic circles and roundabouts generally are safer and more efficient than intersections controlled by stop signs or traffic signals. Two factors account for their safety: 1) the curvature of the vehicle path reduces vehicle speeds, and 2) the likelihood for broadside or head-on collisions is greatly reduced. The capacity of traffic circles and roundabouts exceeds that of intersections controlled by stop signs and traffic signals, because the right-of-way is more efficiently shared. Most entering vehicles do not stop and wait while the right-of-way is assigned to other vehicles (as at a signal), but rather proceed directly into and through the intersection.

Traffic circles and roundabouts are frequently designed as part of a gateway, entry feature, or other landmark feature within residential communities. The center island affords opportunities for landscaping, signs, sculptures, or other types of civic monuments.

Mid-block roundabouts, while not an intersection traffic-control device, are nevertheless an effective traffic-calming device.

FIGURE 3-9
Curb radius is the radius of the circle formed by the curve of the curb at the corner.

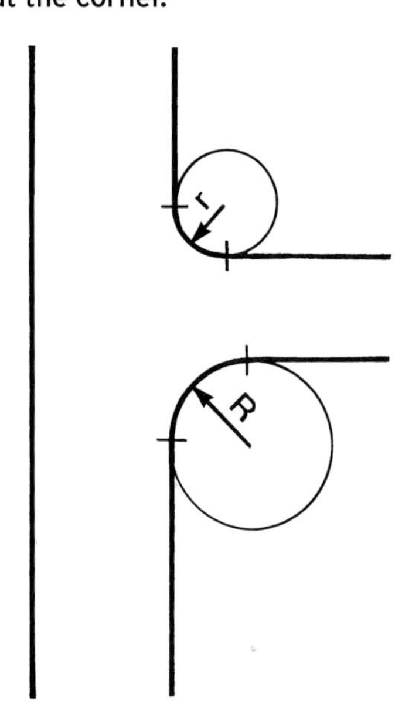

Intersection Spacing

Intersections should be spaced far enough apart so that the traffic stopped to make left turns at one intersection does not back up sufficiently to interfere with traffic movements at the next intersection. Since the spacing required to prevent the backup of left-turning vehicles is dependent upon traffic volumes, trip directions, and turning movements, it is difficult to develop standard recommendations for intersection spacing. On low-volume streets, a distance of 125

GLATTING JACKSON KERCHER ANGLIN LOPEZ RINEHART

FIGURE 3-10
Small curb radius.

feet is usually adequate; on residential collectors, a distance of 250 feet may be more appropriate.

Adequate distance between intersections also eliminates corner cutting, the tendency of drivers to cross two adjacent intersections diagonally via the shortest path, a route that can dangerously traverse the path of opposing vehicles. To eliminate corner cutting, intersections should be spaced a minimum of 125 feet apart (Figures 3-7 and 3-8).

Curb Radius

Curb radius is the radius of the circle joining the intersecting street curbs at a corner (Figure 3-9). The curb radius should accommodate the expected amount and type of traffic and allow for safe turning speeds.

As the curb radius increases, the paving cost and pedestrian crossing distances also increase, dangerous incomplete stops become more frequent, and drivers make turns at higher speeds (Figure 3-10). On the other hand, if the curb radius is inadequate, traffic conflicts tend to rise and vehicles drive over the curbs. Table 3-1 provides recommended ranges for curb radii by type of intersec-

TABLE 3-1

RECOMMENDED RANGES FOR CURB RADII

Type of Intersection	Curb Radius (feet)
Local-Local	10-15
Local-Collector	15-20
Collector-Collector	15-25

Source: American Association of State Highway and Transportation Officials, *A Policy on Geometric Design of Highways and Streets* (Washington, D.C.: AASHTO, 1990).

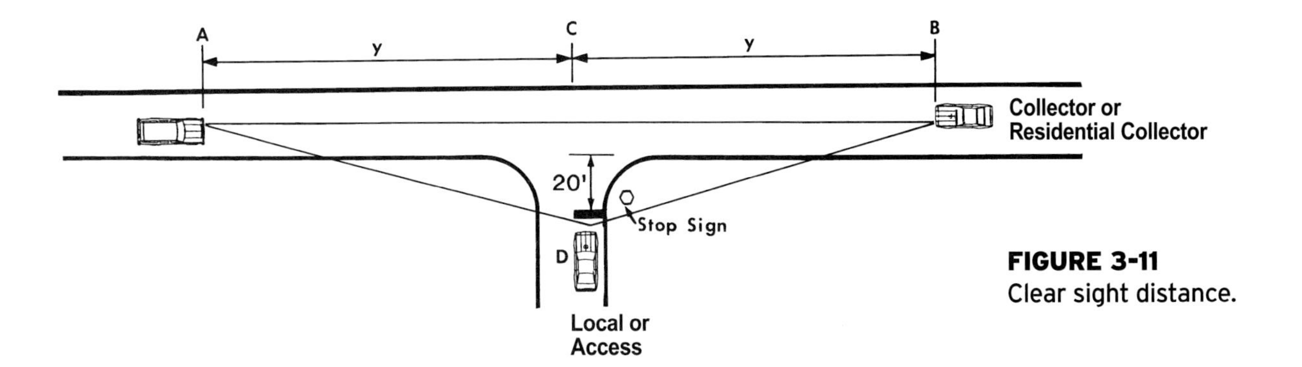

FIGURE 3-11
Clear sight distance.

tion. Local-residential collector intersections and residential collector-residential collector intersections are recommended for higher curb radii based on a desire to ease slightly the steering motion of drivers entering or leaving the collector. However, this additional convenience for drivers comes at the price of longer crossing distances for pedestrians, higher vehicle speeds at points of pedestrian-vehicle conflict, and more pavement.

At most residential street intersections, trucks and other large vehicles making turns should be expected to use the full width of pavement, thereby briefly encroaching on the pavement used by the opposing flow of traffic. Attempts to eliminate this encroachment will result in large curb radii that are highly inappropriate for residential streets.

Vertical Alignment on Intersection Approaches

The intersection and approach areas where vehicles store while waiting to enter intersections should be designed with a relatively flat grade. The grade should generally not exceed 5 percent where practical (AASHTO, 1984). Where ice and snow create hazardous conditions, the desirable grade on the approach leg is

TABLE 3-2

SIGHT DISTANCE AT INTERSECTIONS

Design Speed (mph)	Intersection Sight Distance[1] (feet)
15	105
20	125
25	150
30	200
35	225-250
40	275-325
45	325-400

[1] Distance Y in Figure 3-11.

Source: American Association of State Highway and Transportation Officials, *A Policy on Geometric Design of Highways and Streets* (Washington, D.C.: AASHTO, 1990). For design speeds below 20 mph, distances are extrapolated from AASHTO formulas.

0.5 percent, but should not exceed 2 percent where practical. These grades should be maintained for a minimum of 50 feet from the intersection, and a full 100 feet is more desirable.

Corner Sight Distance

At the intersection of streets with different classifications, traffic on the lower-order functional classification street should be made to stop or yield. (At traffic circles and roundabouts, all entering traffic yields.) Also, the design of the lower-order street should provide an adequate corner sight distance (Table 3-2). Adequate sight distance permits drivers entering the higher-order street to see approaching traffic from a long enough distance to allow them to decide when to enter and to accelerate in advance of the approaching traffic. The entire area of the clear sight triangle should be designed to provide the driver of the entering vehicle (point D in Figure 3-11) with an unobstructed view to all points three feet above the roadway along the centerline from point A to point B. The recommended distance (Y) depends upon the design speed of the higher-order street (Table 3-2).

This clear sight distance does not necessarily apply to both sides of divided parkways, where full stops in the median landing area are likely. A clear sight triangle can be recalculated for one side of a parkway from the median landing area.

4

Streets as Drainage System

While residential streets play several primary roles—providing access to properties, conducting traffic, accommodating bicycles and pedestrians, and creating a visual and social setting for homes—they also serve an important secondary function, which is to collect and convey stormwater runoff. In fact, drainage is a major consideration in street design. In developed areas, the percentage of rainwater that quickly runs off the land is high because of large expanses of impervious surface that prevent water from soaking into the soil. Further, the presence of significant quantities of stormwater runoff can be a potential hazard to life and property.

Closed and Open Systems

In a closed drainage system, runoff is collected and retained within the roadway by curbs and gutters that convey the flow to the main drainage system (Figure 4-1). Where storm sewer mains are available under or near the roadway, the flow is removed at frequent intervals from the street cross-section through storm drains (inlets opening on the curb) that are connected to the storm sewer mains. The three variables in this type of collection system are the amount of flow from adjacent property, the flow capacity of the street section, and the capacity of the curb inlets.

Open drainage systems are an alternative method that encourages infiltration of stormwater into the soil. The sheet flow of water across streets conveys

FIGURE 4-1
Catch basin in closed drainage system.

CLYDE DAVIS. COURTESY OF MOBILE COUNTY (ALABAMA) PUBLIC WORKS

water to swales and adjacent lawns, reducing the need for costly structures (Figure 4-2). A swale is typically a grass-lined open channel adjacent to a street without curbs. The use of swales allows a significant percentage of the stormwater to seep into the earth immediately, while the remainder soaks in during the hours following a rain. Swales reduce stormwater runoff, especially the amount of peak runoff, and help remove some of the pollutants, including motor oil, from the stormwater. Appropriate soils and slopes are required for the proper functioning of open drainage systems, and slopes need to be compatible with soil types. In some cases, paving the swales may be necessary.

FIGURE 4-2
Open drainage systems with swales.

CLYDE DAVIS. COURTESY OF MOBILE COUNTY (ALABAMA) PUBLIC WORKS

BOB DUNPHY

The open drainage approach is compatible with development concepts that organize land uses around path and walkway systems and low-traffic streets with large lots (Figure 4-3). The open system, representing a return to urban practice common in the early part of the 20th century, provides temporary streetside storage for stormwater. By reducing runoff, open drainage can reduce widespread urban flooding. (The proliferation of closed drainage systems has accelerated downstream runoff, creating significant urban flooding problems where none previously existed.) Open drainage may result in occasional ponding of water, and the maintenance of swales may be a consideration in some areas.

Whether a system is open or closed, drainage system planning should parallel street layout and gradient planning, with particular attention devoted to the following:

- The function of streets as part of the stormwater management system.
- Street slopes in relation to stormwater capacity and flow velocity in gutters and street swales.

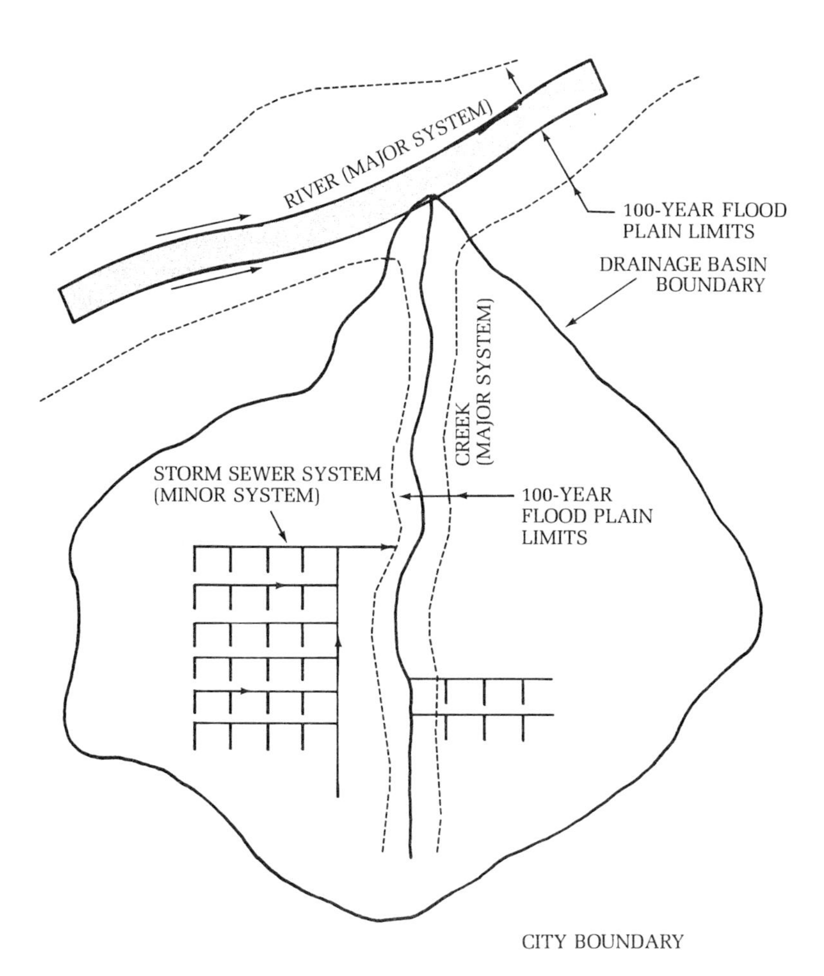

FIGURE 4-3
Open drainage development plan.

- Location and sizing of street culverts. Culverts may be sized to create temporary upstream storage, if earthbank stability and potential overflow effects during major floods are properly considered.
- Location of streets in relation to natural streams, storage ponds, and open channel components of the system.
- Location and capacity of pipe inlet points in relation to gutter slopes, the spread of water across streets, and the flow of water across intersections.
- Coordination of street grades with lot drainage. Positive slope away from all sides of houses is essential. Lot drainage becomes difficult when the fall from the earth grade at the center rear of a house to the street curb at the lowest front corner of the lot is less than 1.5 to 2 percent (usually from 14 to 24 inches).

Street and Curb Cross-Sections

The most commonly used cross-section is a center crown sloping at a rate of one-quarter inch per foot toward a swale or curb and gutter on each side of the street (Figure 4-4). A slope of three-eighths inch per foot is recommended in areas that must accommodate short periods of intensive rainfall. Sidewalks should be separated from the curb by a planting area. On steep sidehill sections, sidewalks may be located against the curb to minimize the grading required for the street and sidewalk.

On sidehill sections, street cross-sections can be designed with a cross-slope or with a crown at the one-third point with a slope toward each curb. This section should not be used on residential collector streets that convey relatively high-speed traffic, as the ridge tends to make control of the car more difficult. In any case, the cross-slope on residential roadway sections should not exceed 5 percent.

Since construction economy often dictates the slipforming of both curbs and integral curbs and gutters, uniform curb and gutter sections should be used as much as possible. The hydraulic capacity of vertical curbs makes these curbs suitable for both side-opening curb inlets and grate-opening inlets. Sloping curbs, on the other hand, have a lower hydraulic capacity and must transition to a vertical face to accommodate side-opening curb inlets. Where limited hydraulic capacity is not a significant issue, the flexibility of sloping curbs to accommodate field changes in driveway locations is a characteristic that recommends them.

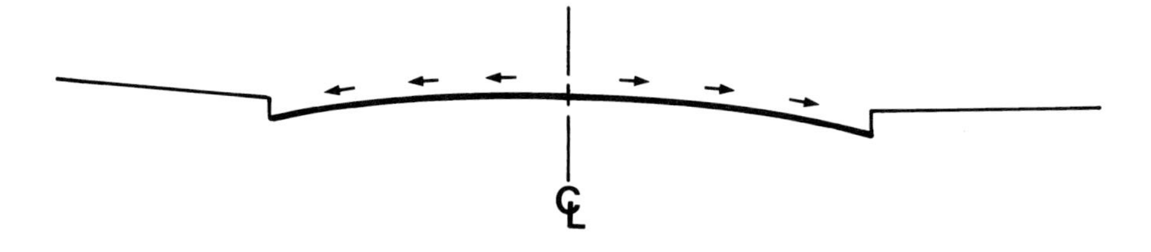

FIGURE 4-4
Cross section of a street showing a center parabolic crown.

Runoff Amounts

Rainfall runoff amounts are typically calculated by what is called the Rational Method, which, in spite of its shortcomings, has gained wide acceptance as the best method currently available. The Rational Method is based on the direct relationship between rainfall and runoff and accounts for both the average intensity of rainfall in an area and the different runoff characteristics associated with various land uses.

During a typical storm, there is a relatively short period when rainfall intensity reaches a peak several times as great as the average intensity over a longer period. It is for this period of peak runoff that stormwater systems must be designed.

The runoff coefficient is an estimate of the proportion of rainwater that strikes a surface and runs off. The runoff coefficient is based on the amount of impervious surface such as roofs, driveways, street paving, and sidewalks associated with urbanization. Lawns also contribute to runoff, although their runoff coefficients are dependent upon soil type and ground slope.

Using the Rational Method, drainage areas are subdivided so that runoff contributed to each gutter or swale can be computed at the end of a block or at other points where an inlet or pickup point is required. Inlets are usually sized so that a portion of the flow is bypassed, sending it along to the next reach of gutter. The flow reaching the second inlet is a portion of the flow contributed from its drainage area plus the flow bypassing the first inlet. The low point (or sump inlet) catches the remaining flow from both directions, and must be sized accordingly.

The locations and required capacities of inlets and swales are established by computing estimated flow rates, depth and velocity of flow, and spread across the street. A properly sized sump inlet may provide a degree of outflow control from the sump storage by allowing the temporary ponding of stormwater. When, during a major storm, the storm drain system is filled beyond its capacity, the excess stormwater flows across lawns and streets.

Criteria for the Spread of Water across Streets

The allowable spread of water across streets from the curb is limited by the requirement to maintain two clear eight-foot moving lanes of traffic for residential collector streets during minor storms. One clear lane should be maintained on local streets, while access streets may have a spread equal to one-half of their width. These criteria may be difficult to justify in arid areas where the gradient is minimal, runoff may contain considerable suspended solids, drainage occurs primarily through surface flow, and the public accepts the inconveniences associated with the spread of water during infrequent storms.

When a steep cross-slope is used (5 percent is suggested as a maximum), a ten-foot spread would produce an excessive depth at the curb, making it difficult to intercept the flow except with a particularly long curb-opening inlet. Consequently, if pavement is cross-sloped from 3 to 5 percent, the depth of flow at the curb should not exceed six inches.

Flow across Intersections

The intersection of a street on a grade with another street, especially a residential collector or local street, creates a critical situation. Even when the flow on the grade is severely limited, great care must be taken to provide inlets that will intercept virtually all the flow from a minor storm. Full interception of the flow from a major storm may be impossible. A T-intersection requires special care, because houses directly below the T-stem are particularly subject to damage from intersection overflow during major storms. The overflow of T-intersections can be somewhat impeded by installing a higher-than-normal roadway crown on the through street, adding additional inlets, and providing diversion swales between residences.

Unrestricted flow across residential collector streets should not be allowed during frequent storms. Although controlled flow across local streets is acceptable, there are no design limits placed on the flow across access streets. During major storms, permissible flow across intersections is a function of the street's traffic-carrying importance and the availability of convenient alternative routes.

Ponding at Low Points in Grade

Even during a minor storm, some ponding of water at low points in the grade is inevitable and, in some cases, may even be desirable when the expense of constructing an additional inlet is a factor. To reduce ponding, street design must permit most of the flow to be intercepted before it reaches the bottom of the grade. Interception of most of the flow before deposition at the sump permits removal of a large percentage of the sediment carried by the flow and thereby prevents sediment accumulation in the gutter as velocity along the decreasing grade slows. A curb-opening inlet at the low point in the grade is an efficient structure that must be amply sized to handle the likely accumulation of debris. Engineers can ensure effective performance during a major storm by considering damage from water overtopping curb and sidewalk, methods of minimizing damage, relative sizing of inlet and pipe, and overflow mechanisms.

Maximum Velocity in Gutters

Water flowing down steep grades can be dangerous. For example, water cascading at ten feet per second can exert a force of 100 pounds against a flat, one-foot-wide object placed across the flow. A 20-pound push at shoe level can sweep adults off their feet. Obviously, gutter flow can present a serious hazard to children. In addition, rapid gutter flows are difficult to intercept at inlets and can result in water either shooting across an intersecting street or overriding curbs and causing severe localized erosion and sometimes damage to downhill properties.

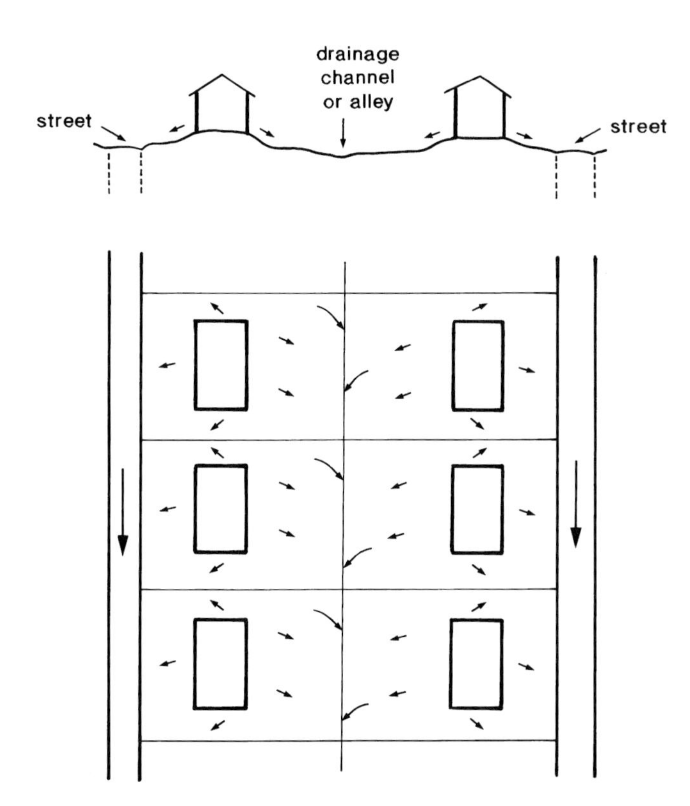

FIGURE 4-5
Block and lot grading. Arrows show the flow of water away from houses and into streets and drainage channel or alley (at left). The slope of lots on flat land can be increased by proper grading (at right).

If the calculated velocity of water in the deepest part of the gutter exceeds ten feet per second, designers must reduce the allowable discharge into the gutter until the velocity falls below this limit. Figuring out where and how to reduce the runoff entering the gutter can be a problem. One expensive solution would be to install additional inlets upstream. Alternatively, engineers might try to divert runoff to a path other than the steep street, preferably through revisions to the street design layout. Street resurfacing plans that reduce capacity should be considered in the calculations.

Block and Lot Grading

Proper grading is an important element in preventing wet basements, damp crawl spaces, eroding stream banks, muddy yards, and overflowing septic tank systems. It also eliminates costly corrective work such as retaining walls, regrading operations, and extra drainage-pipe lines. The planners responsible for determining the key grade elevations and the construction superintendent and grading foreman responsible for overseeing on-site work must have the necessary knowledge to achieve construction savings and proper grading improvements.

The planning and execution of good grading involves certain basic steps pertaining to street layout and block and lot grading (Figure 4-5). The objective is to establish the street grades, floor elevations, and lot grades in proper relation to each other and to the existing topography, while paying attention to considerations of the property's appeal and use.

Flat land represents the most difficult type of terrain on which to create positive surface drainage away from lots and blocks and buildings to streets or other suitable outlets. The problems associated with grading flat terrain stem mostly from foundation heights not sufficiently above existing ground levels and the need for fill material. The slope of lots on flat land can be increased for improved surface drainage by lowering the street profile. Such a profile modification not only lowers the outfall of swales and driveways but also provides fill material for the abutting lots with practically no soil surplus. If paved driveways are used for drainage channels at relatively flat grades, rather than grass swales at steeper grades, the gradient of grass swales may be reduced below the gradient normally used. Finally, drainageways may be used along rear lot lines. Another solution to the need for fill material is to construct houses with basements rather than slabs.

Cut-and-Fill Embankment Slopes

The economic significance of slope instability is much greater along residential streets than along rural highways. Highway slope failures generally involve a few hours of inconvenience for highway users and a moderate maintenance outlay, whereas slope failures in developed residential areas generally involve utilities, street improvements, and, often, buildings—all of which are more costly to restore and the restoration of which occasions considerable inconvenience of the public.

Cut-and-fill slopes within rights-of-way should normally be designed for a maximum slope no steeper than 3:1, and a more gradual 4:1 slope is preferable. Where a slope easement cannot be obtained and a slope steeper than 3:1 must be used, a four- or five-inch thickness of concrete riprap is recommended. Vertical cut-and-fill embankment slopes require retaining walls.

The kinds of situation in which slope or embankment stability may become a problem are generally identifiable within any geographic area. Local geologic engineering consultants can provide appropriate guidance on identifying potential stability hazards as well as practical methods for their mitigation.

5

Pavement

Although paved streets are taken for granted in the United States today, there was a time when thousands of miles of unpaved roads and city streets handled substantial amounts of traffic. The alternately dusty and muddy conditions that worsened with increased traffic volumes were an important impetus for the adoption of subdivision regulations that require the paving of new streets. In addition to controlling the dust and mud problem, a well-designed and -constructed pavement confers benefits such as reduced soil erosion, a safer traveling surface for vehicles, and the capacity to handle loads without sinking or shifting. Cost, of course, is an important factor in choice of pavement. Both the initial cost and the cost of maintenance should be considered, as should the aesthetics of the wearing surface.

Pavement for residential streets must be designed for the volume and characteristics of the traffic expected to use the streets (Figure 5-1). Just as residential street standards do not apply to arterials, standards for state and county arterial and major collector roads are not appropriate for residential streets. Likewise, pavement designers may find it useful to distinguish collector streets from lower-order streets, since collectors are more likely to carry heavy loads, including buses.

The field of pavement design is dynamic. Concepts are continually changing as new data become available. The broad range of street design methods reflects variations in local opinion regarding the suitability of various pavement options.

FIGURE 5-1
Pavement for residential streets must be designed for the volume and characteristics of the traffic expected to use the streets.

FISHER ISLAND

THE WALT DISNEY COMPANY

In particular, materials available for the construction of pavements have a major influence on design. Nonetheless, a common set of design principles applies to all situations regardless of individual circumstances. The design of all streets must account for the soil (subgrade), the available range of paving materials, and the behavior of those materials under load and in all probable climactic conditions.

Communities should consider the wide range of pavement options for residential streets when determining their requirements on pavement types. Pavement standards have been developed for different types of materials, and are usually a part of a locality's subdivision, site planning, or planning or public works agency design requirements. Pavement standards should be based on the climate and soil characteristics of the local area.

Many pavement trade associations—including the Asphalt Institute, the Portland Cement Association, the National Stone Association, and the National Lime Association—have developed standards for residential street pavements. Many localities have adopted some of these standards or have developed their own. Often, local conditions such as subgrade soils and available construction materials are the basis for design variations from region to region. In addition, a community's aesthetic values relating to such factors as the color and texture of street surfaces as well as local tradition usually play an important role in the selection of pavement material.

The components of street pavement design include the subgrade, underground drainage, the base, and the wearing surface.

Subgrade

All pavements derive their ultimate support from the underlying subgrade. The subgrade is the foundation layer for the street. It may be simply the natural earth surface or compacted soil, or it may include additives for stabilizing the soil.

Preparation of the subgrade is one of the most important steps in street construction. The subgrade soil must be of sufficient load-bearing capacity as determined through laboratory tests. If laboratory test equipment is not available, street design may be based upon a careful field evaluation—visual inspection and simple field tests—performed by an experienced soils engineer.

Substandard soils can be stabilized by incorporating an additive, such as lime, into the soil. Such additives are used especially for clay soils. Cement or fly-ash can be mixed with a substandard soil to stabilize it, or a layer of crushed stone or the equivalent can be installed to improve the load-carrying capability of poor subgrades.

Base

The base course is the layer of material that lies immediately below the wearing surface of a pavement, and there may be also a subbase layer of material between the base and subgrade. Base courses may be constructed of graded aggregates, slag, soil/aggregate mixtures, cement-treated granular materials, bituminous aggregate mixtures of several types, or econocrete (a low-strength—750 psi—concrete base).

The function of the base course varies according to the type of pavement, but, in general, base courses are used for 1) protection against frost action, 2) drainage, 3) prevention of volume change of the subgrade, 4) increased structural capacity, and 5) expedition of construction.

To provide drainage, the base may or may not be a well-graded material, but it should contain little or no fines (sandy clays) and have good permeability. A base course designed for frost action should be non-frost susceptible and free draining with no plastic fines. A base course need not be free draining to provide structural capacity, but it should be well graded and should resist deformation due to loading. To provide resistance to deformation, it is often necessary to stabilize the base course with cement or asphalt, or to install lean concrete or econocrete.

Wearing Surface

Historically, pavements have been divided into two broad categories—rigid and flexible, or concrete and asphalt. The major difference between the two types of pavements is the manner in which they distribute the load over the subgrade. Rigid pavements tend to distribute the load over a relatively wide area of soil. Thus, a major portion of the structural capacity is supplied by the slab itself. The chief factor in the design of rigid pavement is the structural strength of the concrete. As a result, minor variations in the strength of the subgrade have little influence on the structural capacity of the pavement.

By contrast, flexible pavements maintain direct contact with and distribute loads to the subgrade, depending upon aggregate interlock, particle friction, and cohesion for stability. Flexible pavements have a layer (or layers) of aggregate that provides the aggregate interlock and particle friction, topped by a layer (or layers) of aggregate bound together with asphalt, which provides the necessary cohesion. Both rigid and flexible pavement sections, if properly designed, can be placed on the raw subgrade.

Asphalt

Liquid asphalt is obtained as a residue during the petroleum refining process. By far the most common pavement material in use is hot mix asphalt. It is made in a plant under controlled conditions by mixing liquid asphalt cement with accurately proportioned aggregates. After the material is transported to the site,

TORTI GALLIS

FIGURE 5-2
Asphalt was used for the street surface at South Riding in Loudoun County, Virginia.

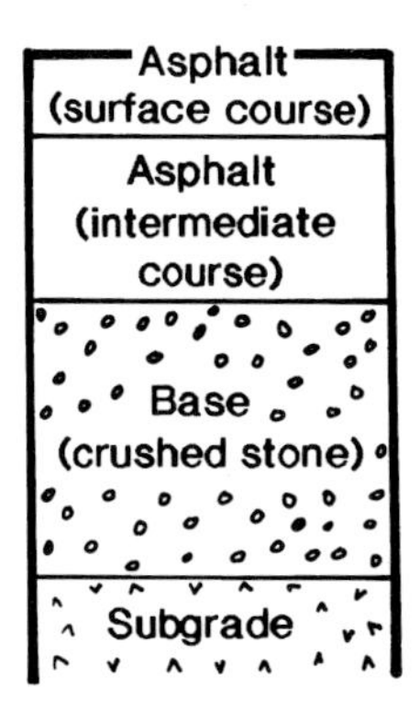

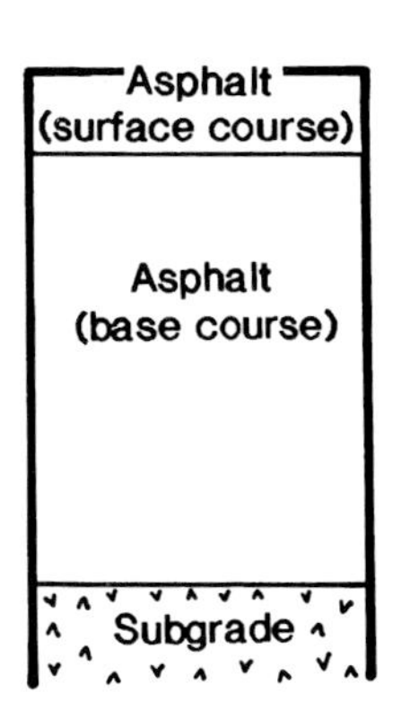

FIGURE 5-3
Typical asphalt street structures.

specially designed paving machines install the mixture, while it is still hot, to the required thickness and grade specifications, overlaying the prepared subgrade. After the mixture has been rolled to the desired compaction, the surface is ready for immediate use by traffic (Figure 5-2).

Putting an asphalt surface on a base of crushed stone is the traditional method of building hot mix asphalt streets. Another accepted method, called full-depth hot asphalt, is to place the asphalt directly on the compacted earth. Whichever method is used, the asphalt paving often is applied in two layers—a base layer and a surface layer (Figure 5-3). When developing a subdivision, the base layer can be installed early in the development process and used to handle the heavy traffic from construction vehicles. After the houses are completed, the developer can install the top layer so that a clean smooth surface is presented to the homebuyers. Information on asphalt construction procedures is available from the National Asphalt Paving Association.

Concrete

The most common form of concrete used in street construction is made from Portland cement, an extremely fine powder manufactured in a cement plant. When mixed with water, Portland cement forms a paste that binds materials such as sand and gravel or crushed stone into concrete. As with asphalt streets, concrete streets require careful preparation of the subgrade, including uniform compaction and, in some cases, the installation of a base layer of crushed stone, sand, or soil-cement (Figure 5-4).

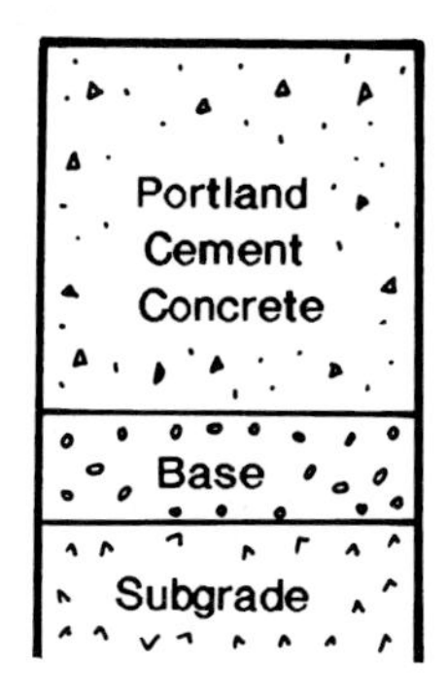

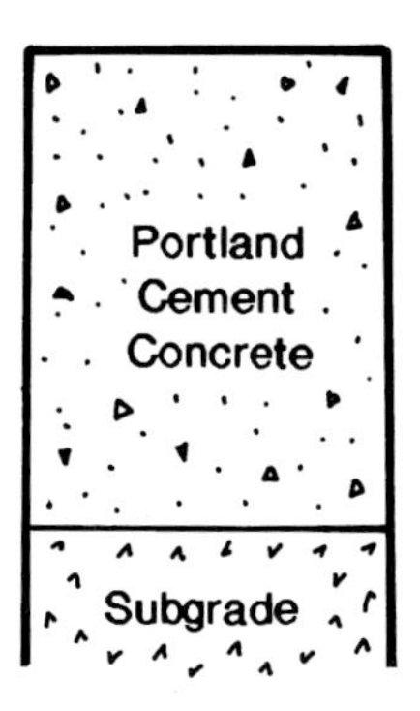

FIGURE 5-4
Typical concrete street structures.

FIGURE 5-5
Concrete street.

BOB DUNPHY

The most common method of constructing concrete streets (Figure 5-5) is slip-form paving with integral curb. A slip-form paver is set to line and grade and the concrete is spread, consolidated, screeded, and finished in one pass. Contraction joints are then sawed at traverse intervals not to exceed in feet twice the thickness in inches. (For example, the joint spacing for a six-inch-thick residential street would be 12 feet.) Information on concrete construction procedures is available from the American Concrete Pavement Association.

Life-Cycle Cost Analysis

Communities planning residential street schemes should be prepared to measure the economic impact of alternative roadbuilding strategies, particularly when funding sources are limited. Economic impact is best determined by conducting some form of discounted cash flow analysis that recognizes the timing and size of all financial outlays associated with the project.

Typically, the life cycle for highways includes the construction and maintenance needed to keep investment integrity (serviceability) within acceptable standards up to the point where the first-year actions (new construction) should be repeated. The life cycle for arterial and major collector roads includes construction, maintenance, rehabilitation, maintenance, and, finally, salvage costs before cycling back to reconstruction (AASHTO, 1978). For residential streets, the absence of heavy traffic results in an especially long pavement life that does not fall within normal time horizons for highway life-cycle analysis. The extended life cycle of residential streets causes no methodological problems for analysis, though expenditures 30 years in the future have small discounted values, but such information may need to be carefully presented to local agencies whose planning horizons are much shorter.

References

American Association of State Highway and Transportation Officials. *Manual on User Benefit Analysis of Highway and Bus-Transit Improvement.* Washington, D.C.: AASHTO, 1978.

——. *A Policy on Geometric Design of Highways and Streets* (a.k.a. the AASHTO Greenbook). Washington, D.C.: AASHTO, 1990 (new edition forthcoming 2001).

Bucks County Planning Commission. *Performance Streets.* Doylestown, Pennsylvania: Bucks County Planning Commission, 1980.

California Department of Transportation. *California Highway Design Manual.* Fifth Edition, Chapter 1000, "Bikeway Planning and Design." Sacramento, California: California Department of Transportation, February 2001.

Institute of Transportation Engineers. *Guidelines for Residential Subdivision Street Design: A Recommended Practice.* Washington, D.C.: Institute of Transportation Engineers, 1993.

——. *Traditional Neighborhood Development Street Design Guidelines: Recommended Practice.* Washington, D.C.: Institute of Transportation Engineers, 1999.

——. *Traffic Engineering for Neo-Traditional Neighborhood Design: An Informational Report.* Washington, D.C.: Institute of Transportation Engineers, 1994.

——. *Trip Generation.* Sixth Edition. Washington, D.C.: Institute of Transportation Engineers, 1997.

Kerr, William O., and Barbara A. Ryan, *Avoiding the Pitfalls of Life-Cycle Cost Analysis.* Washington, D.C.: Washington Economic Research Consultants, October 1987.

Marks, Harold. *Traffic Circulation Planning for Communities.* Los Angeles: Gruen Associates, 1974.

Pennsylvania Department of Transportation. *Guidelines for Design of Local Roads and Streets.* Harrisburg, Pennsylvania: Pennsylvania Department of Transportation, 1983.

ULI–the Urban Land Institute. *Project Infrastructure Development Handbook.* Washington, D.C.: ULI–the Urban Land Institute, 1989.

ULI–the Urban Land Institute, American Society of Civil Engineers, and National Association of Home Builders. *Residential Streets: Objectives, Principles & Design Considerations.* Washington, D.C.: ULI–the Urban Land Institute, American Society of Civil Engineers, and National Association of Home Builders, 1974.

U.S. Architectural and Transportation Barriers Compliance Board, *Accessible Rights-of-Way: a Design Guide*. Washington, D.C.; U.S. Architectural and Transportation Barriers Compliance Board, 1999.

U.S. Department of Transportation, Federal Highway Administration. *Manual of Uniform Traffic Control Devices for Streets and Highways*. Washington, D.C.: U.S. Department of Transportation, 2000.

Wright, Paul H., and Norman J. Ashford. *Transportation Engineering: Planning and Design*. Third Edition. New York: John Wiley & Sons, 1989.

Index

AASHTO Greenbook, 6, 11, 13, 14, 26, 30
Accessibility, 42, 44
Additives, 69
ADT (Average daily traffic), 14, 16
Aggregates, 69
Alignment, 19, 21, 31–32
Alleys, 28-29
American Association of State Highway and Transportation Officials (AASHTO), 6, 11, 13, 14, 44, 57, 72
American Concrete Pavement Association, 72
Americans with Disabilities Act (ADA), 44–46
Arterial streets, 10–11, 16, 19, 44
Asphalt, 70–71
Asphalt curbs, 48, 49
Auto court, 35–36
Average daily traffic (ADT), 14, 16

Base, 69–71
Basements, 66
Bicycle travel, 12, 19, 20, 25, 40, 42–44
Block grading, 65–66
Bucks County (Pennsylvania) Planning Commission, 5

Chicanes, 50
Circular length, 33–36
Circular turnarounds, 32–34
Circulars, 32–34
Classification of streets, 9–13
Clear sight distance, 57–58
Closed drainage system, 59–62
Cluster development, 3
Collector intersections, 57
Collector streets, 11–12, 14, 16, 22, 25, 26, 42, 44, 64, 72
Community planning, 7, 68
Concrete, 69, 70, 71–72
"Corner cutting," 44–46
Cross-sections, 62
Crosswalks, 45–46
Crushed stone, 69, 71
Culverts, 62
Curb radius, 56–57
Curbs, 44-49, 63
Curves, 31, 32
Curvilinear streets, 19, 21
Cut-and-fill embankment slopes, 66

Dead end streets, 32–36
Delivery vehicles, 15, 27, 34, 55
Drainage, 18, 25, 30, 47, 48, 59–66
Driveways, 27, 28, 36–37, 41, 47, 66

Emergency vehicles, 14–15, 17, 27, 34
Entrances to residential communities, 17–18, 55
"Eyebrow," 36

Federal Highway Administration, 39
Federal Housing Administration (FHA), 4
Fire trucks. See Emergency vehicles
Flooding, 61, 62
Free flow, 13
Freeways, 10, 16, 72

Gradients, 30–31
Grading, 25, 65–66
Grassy strips, 26, 41
Grid layout, 18–19
Gutters, 61, 62, 64–65

Hierarchy of street uses, 7, 9, 21, 67
Hilly terrain, 30–32
History of residential streets, 1–3
Horizontal curves, 32

Institute of Transportation Engineers (ITE), 5, 17, 20, 21
Intersection angle, 53
Intersection spacing, 55–56
Intersections, 21, 51–58, 64

Lanes, number of, 21–22
Layout of streets, 18–21
Life-cycle cost analysis, 72
Local streets, 12, 14, 16, 21, 22, 26, 64
Lot grading, 65–66
Lot width, 21, 28, 41

Manual on Uniform Traffic Control Devices (M.U.T.C.D.), 39–40
Medians, 22

National Asphalt Paving Association, 71
Neotraditional development, 19–20

Open drainage systems, 59–62
Overdesign, 7

Parking, 15, 16, 22, 25, 26–28, 36, 47, 48, 49
Paths, 42
Pavement, 67–72
Pavement width, 22–25
Pedestrians, 12, 19, 20, 31, 40–42
Performance Streets (Bucks County, Pennsylvania), 5, 6–7
Planned unit development (PUD), 3
Platting, 18–21
Ponding, 30, 61, 64
Portland cement, 71
Private streets, 8
Public transit, 21, 67

Rational Method to calculate runoff, 63
Raveling, 49
Right-of-way, 25–26, 41
Roundabouts, 53–55
Runaways, 47
Runoff, 18, 27, 59–60, 63

Security, 17, 18
Shared driveways, 36–37
Sidewalks, 25, 26, 40–41, 44–49, 62
Signs, 29, 39–40, 44
Slopes, 61–62, 66
Sloping curbs, 47–49
Slow flow, 13
Snow plows, 15, 17, 25, 27, 41, 47
Soils, 18, 69
Solid waste collection, 15, 34, 41
Speed, 15-16, 29–30
Splitter islands, 54
Spread of water, 63
Storm sewers, 59
Stormwater, 27, 29, 48, 59–63
Streetscape, 37–40
Subcollector, 57
Subgrade, 69, 70, 71
Sump inlets, 63, 64
Swales, 60-62, 64, 66

Textured pavement, 46, 50
Topographic mapping, 18
Traffic calming, 49–50, 55
Traffic circles, 53–55
Traffic flow, 13–14
Traffic volume, 15–17
Trees, 25, 26, 27, 32, 37–39
Trip generation, 16–17
Turnarounds, 32–36

U.S. Department of Housing and Urban Development (HUD), 6
Utilities, 15, 25, 39, 41, 66

Vegetation, 37–39
Vertical alignment, 31–32, 57–58
Vertical curbs, 46–47
Wearing surface, 67, 70–72
Widening, 25
Width (of pavement), 22–25
Width (of right-of-way), 21, 25–26

Yield flow, 14

Zoning, 3